Communications
in Computer and Information Science   1862

Rationale

The CCIS series is devoted to the publication of proceedings of computer science conferences. Its aim is to efficiently disseminate original research results in informatics in printed and electronic form. While the focus is on publication of peer-reviewed full papers presenting mature work, inclusion of reviewed short papers reporting on work in progress is welcome, too. Besides globally relevant meetings with internationally representative program committees guaranteeing a strict peer-reviewing and paper selection process, conferences run by societies or of high regional or national relevance are also considered for publication.

Topics

The topical scope of CCIS spans the entire spectrum of informatics ranging from foundational topics in the theory of computing to information and communications science and technology and a broad variety of interdisciplinary application fields.

Information for Volume Editors and Authors

Publication in CCIS is free of charge. No royalties are paid, however, we offer registered conference participants temporary free access to the online version of the conference proceedings on SpringerLink (http://link.springer.com) by means of an http referrer from the conference website and/or a number of complimentary printed copies, as specified in the official acceptance email of the event.

CCIS proceedings can be published in time for distribution at conferences or as post-proceedings, and delivered in the form of printed books and/or electronically as USBs and/or e-content licenses for accessing proceedings at SpringerLink. Furthermore, CCIS proceedings are included in the CCIS electronic book series hosted in the SpringerLink digital library at http://link.springer.com/bookseries/7899. Conferences publishing in CCIS are allowed to use Online Conference Service (OCS) for managing the whole proceedings lifecycle (from submission and reviewing to preparing for publication) free of charge.

Publication process

The language of publication is exclusively English. Authors publishing in CCIS have to sign the Springer CCIS copyright transfer form, however, they are free to use their material published in CCIS for substantially changed, more elaborate subsequent publications elsewhere. For the preparation of the camera-ready papers/files, authors have to strictly adhere to the Springer CCIS Authors' Instructions and are strongly encouraged to use the CCIS LaTeX style files or templates.

Abstracting/Indexing

CCIS is abstracted/indexed in DBLP, Google Scholar, EI-Compendex, Mathematical Reviews, SCImago, Scopus. CCIS volumes are also submitted for the inclusion in ISI Proceedings.

How to start

To start the evaluation of your proposal for inclusion in the CCIS series, please send an e-mail to ccis@springer.com.

Henri Emil Van Rensburg · Dirk Petrus Snyman ·
Lynette Drevin · Günther Richard Drevin
Editors

ICT Education

52nd Annual Conference of the Southern African
Computer Lecturers' Association, SACLA 2023
Gauteng, South Africa, July 19–21, 2023
Revised Selected Papers

Springer

Editors
Henri Emil Van Rensburg 🆔
North-West University
Potchefstroom, South Africa

Dirk Petrus Snyman 🆔
University of Cape Town
Rondebosch, South Africa

Lynette Drevin 🆔
North-West University
Potchefstroom, South Africa

Günther Richard Drevin 🆔
North-West University
Potchefstroom, South Africa

ISSN 1865-0929   ISSN 1865-0937 (electronic)
Communications in Computer and Information Science
ISBN 978-3-031-48535-0   ISBN 978-3-031-48536-7 (eBook)
https://doi.org/10.1007/978-3-031-48536-7

This Springer imprint is published by the registered company Springer Nature Switzerland AG
The registered company address is: Gewerbestrasse 11, 6330 Cham, Switzerland

Paper in this product is recyclable.

Preface

The 52nd Annual Conference of the Southern African Computer Lecturers' Association was held from July 19th to 21st, 2023, at the serene 26 Degrees South in Muldersdrift, South Africa, organised by the School of Computer Science and Information Systems, North-West University, South Africa. The event was co-located with the annual meeting of the South African Institute for Computer Scientists and Information Technologists (SAICSIT) for the second year running, thereby giving the delegates the opportunity to attend both events as there is a considerable overlap in the communities.

The theme of SACLA 2023, "Teach the Future: CS, IS, & IT Education in a Changing World," resonated deeply with the challenges and opportunities facing educators in today's fast-paced technological landscape. Over these three insightful days, we engaged in profound discussions, shared experiences, and embraced innovative approaches to ensure that our students are well-prepared for the ever-evolving demands of the digital age. Of note was the number of papers that addressed the challenges and opportunities afforded by the increasingly common use of large language models such as ChatGPT in the higher education landscape.

The international program committee comprised 44 members. A rigorous double-blind peer review process was followed for all submissions. Each submission was reviewed by at least three members of the program committee. This volume presents a collection of selected papers from the conference, representing the top twelve submissions, based on reviewer ratings. This represents an acceptance rate of 29% for this volume from the 42 papers received. Reviewer feedback was provided to the authors, and they were requested to submit a change log and rebuttal to the program committee to indicate how any issues that were raised by the reviewers were addressed. A best paper award was presented to Cheng-Wen Huang, Max Coleman, Daniela Gachago, and Jean-Paul Van Belle for their paper entitled "*Using ChatGPT to Encourage Critical AI Literacy Skills and for Assessment in Higher Education*". The best paper was once more identified based on the ratings that were given by the reviewers.

In addition to the paper presentations, Jacob Spoelstra, Principal Data Scientist at Microsoft, was invited by the organising committee to give a keynote talk on "*ChatGPT demystified*" in which he elucidated the inner workings of ChatGPT and the underlying components of the large language model on which it is built. These include word embeddings, artificial neural networks, and attention mechanisms, shedding light on how they contribute to its exceptional performance. He further discussed the training process and the alignment of the models, based on human feedback. His talk equipped delegates to better understand this emerging technology and its implications in teaching and learning, and research. Furthermore, a workshop on curriculum content and challenges was conducted. Delegates were encouraged to share challenges and ideas in smaller groups, based on similar course content. Feedback on the commonly identified themes was presented during the annual general meeting of the association.

The organising committee would like to thank all the participants, including speakers, delegates, and reviewers for their contributions to a successful SACLA 2023 conference.

Finally, we wish to acknowledge the EasyChair conference management system, which was used for managing the submissions and reviews of SACLA 2023 papers. As for the preparation of this volume, we sincerely thank our publisher Springer for their assistance.

July 2023

Henri Emil Van Rensburg
Dirk Petrus Snyman
Lynette Drevin
Günther Richard Drevin

Organization

Executive Committee

Conference Chair

Lynette Drevin   North-West University (Potchefstroom Campus), South Africa

SACLA President

Estelle Taylor   North-West University (Potchefstroom Campus), South Africa

Program Committee Chairs

Henri van Rensburg   North-West University (Potchefstroom Campus), South Africa

Dirk Snyman   University of Cape Town, South Africa

Gunther Drevin   North-West University (Potchefstroom Campus), South Africa

Program Committee

Shaun Bangay   Deakin University, Australia

Esiefarienrhe Bukohwo   North-West University (Mafikeng Campus), South Africa

Lance Bunt   North-West University (Vanderbijlpark Campus), South Africa

Alan Chalmers   University of Warwick, England

Tendani Chimboza   University of Cape Town, South Africa

Marijke Coetzee   North-West University (Potchefstroom Campus), South Africa

Nosipho Dladlu   North-West University (Mafikeng Campus), South Africa

Tiny Du Toit   North-West University (Potchefstroom Campus), South Africa

Olaiya Folorunsho   Federal University Oye Ekiti, Nigeria

Lynn Futcher   Nelson Mandela University, South Africa

Leila Goosen	University of South Africa, South Africa
Irene Govender	University of KwaZulu-Natal, South Africa
Ruber Hernández-García	Universidad Católica del Maule, Chile
Barry Irwin	Noroff University College, Norway
Bassey Isong	North-West University (Mafikeng Campus), South Africa
Moemi Joseph	North-West University (Mafikeng Campus), South Africa
Christos Kalloniatis	University of the Aegean, Greece
Dimitrius Keykaan	North-West University (Potchefstroom Campus), South Africa
Eduan Kotzé	University of the Free State, South Africa
Wai Sze Leung	University of Johannesburg, South Africa
Janet Liebenberg	North-West University (Potchefstroom Campus), South Africa
Hugo Lotriet	University of South Africa, South Africa
Vusumuzi Malele	North-West University (Vanderbijlpark Campus), South Africa
Liezel Nel	University of the Free State, South Africa
Koketso Ntshabele	North-West University (Mafikeng Campus), South Africa
Jaco Pretorius	North-West University (Vanderbijlpark Campus), South Africa
Helen Purchase	Monash University, Australia
Linda Redelinghuys	North-West University (Potchefstroom Campus), South Africa
Rayne Reid	Noroff University College, Norway
Sebopelo Rodney	North-West University (Potchefstroom Campus), South Africa
Aslam Safla	University of Cape Town, South Africa
Rudi Serfontein	North-West University (Potchefstroom Campus), South Africa
Imelda Smit	North-West University (Vanderbijlpark Campus), South Africa
Hussein Suleman	University of Cape Town, South Africa
Estelle Taylor	North-West University (Potchefstroom Campus), South Africa
Alfredo Terzoli	Rhodes University, South Africa
Walter Uys	University of Cape Town, South Africa
Thomas van der Merwe	University of South Africa, South Africa
Charles van der Vyver	North-West University (Vanderbijlpark Campus), South Africa
Corné Van Staden	University of South Africa, South Africa

Contents

Beyond the Classroom

Student Centered Teaching and Learning

A Framework for Creating Virtual Escape Rooms to Teach Computational Thinking

Theané Janse van Rensburg and Machdel Matthee[✉] [iD]

Department of Informatics, University of Pretoria, Pretoria, South Africa
machdel.matthee@up.ac.za

Abstract. Computational Thinking is considered one of the essential skills for success in the future workplace. However, integrating computational thinking into the curriculum remains an educational challenge. Escape room games could potentially aid in the development of computational thinking skills because they immerse learners in a gamified, problem-solving scenario. This paper describes the development of a framework, using Design Based Research, to guide the implementation of a virtual escape room to teach computational thinking in higher education. The components include understanding the participants and computational thinking learning objectives, choosing the digital platform, deciding on a theme and puzzles and finally evaluating the escape room. The framework was implemented and evaluated in a first-year Information Systems programming setting. Although the findings do not show a significant increase in understanding of computational thinking, participants indicated that the experience with the escape room increased their motivation to learn more about computational thinking. It is therefore suggested that virtual escape rooms be used in addition to other learning interventions for the ultimate computational thinking learning experience.

Keywords: Computational thinking · game-based learning · virtual escape rooms

1 Introduction

Although there are many definitions for computational thinking (CT), many researchers accept Wing's (2006) definition, as an approach to solving problems, designing systems, and understanding human behavior by drawing on concepts fundamental to computer science. Wing (2006) promotes CT as a vital skill for the future, equating its importance to reading, writing, and basic arithmetic. She goes beyond tertiary education to state that CT should be added to every child's analytical ability. As such, a growing number of educators are bringing computational thinking to the core of many disciplines (Mohaghegh and McCauley, 2016; Lamprou and Repenning, 2018). At the center of attention are questions about how this skill can be effectively taught (Mohaghegh and McCauley, 2016; Lamprou and Repenning, 2018). Various research studies have indicated that since game-based learning involves problem-solving, it can also foster CT thinking skills (Tatar and Eseryel, 2019; Durak et al., 2017; Connolly et al., 2008).

H. E. Van Rensburg et al. (Eds.): SACLA 2023, CCIS 1862, pp. 3–17, 2024.
https://doi.org/10.1007/978-3-031-48536-7_1

In recent years, the potential of game-based learning to cater for the needs of the "Net Generation" has captured the interest of researchers and practitioners, leading to its widespread recognition (Plass et al., 2015; Akour et al., 2020; Ding et al., 2017). It is defined as a learning approach that emphasizes how games can be used during teaching and learning (Zaibon and Shiratuddin, 2010). In recent years, one particular type of game-based and gamification tool that has gained tremendous popularity due to its potential to promote the learning of any chosen subject for a diverse group of learners: escape room games (Menon et al., 2019). Nicholson (2018) defines an escape room as a live-action adventure game where players find themselves locked in a room or series of rooms, from which they must escape within a limited time by solving a series of puzzles. According to a study conducted by Menon et al (2019), escape games have the potential to develop CT skills. Various initiatives have been created to support the development of CT through educational escape rooms; however, most initiatives are aimed at unplugged activities (Apostolellis and Stewart, 2014; Berland and Lee, 2011; Wang et al., 2011; Kazimoglu et al., 2012). Very little research has been carried out to explore the impact of virtual escape rooms on the facilitation of CT skills and whether using virtual escape rooms is a suitable method to teach computational thinking. This study will explore how virtual escape rooms can be used as a teaching method for computational thinking by first developing a framework for its implementation and secondly evaluating its impact on CT learning.

2 Literature Review

Throughout the years' various definitions of computational thinking have been suggested; however, researchers have not yet reached a consensus on what the term entails. However, most researchers agree that CT is a thought process that utilizes concepts and thinking skills fundamental to computer science and has the following fundamental components: abstraction, decomposition, generalisation and algorithmic thinking (Lamprou and Repenning, 2018; Selby and Woollard, 2013; Barr and Stephenson, 2011; Angeli et al., 2016; Wing, 2008; Wing, 2006; Wing, 2011). These components are described in Sect. 2.1.

2.1 Components of Computational Thinking

- Decomposition is the process of breaking down a problem into smaller and more manageable sub-problems.
- Abstraction is the process of simplifying a problem by reducing the unnecessary detail and number of variables to create a more straightforward solution.
- Generalisation is the process of formulating a solution in a generic way so that it can be applied to different problems even though the variables are different.
- Algorithmic thinking is the process of constructing a series of ordered steps that may be followed to provide solutions to problems.

2.2 Integrating Computational Thinking into the Curriculum

In a literature review conducted by Hsu et al. (2018), researchers have reported on the following strategies to teach CT: problem-based learning, project-based learning, collaborative learning, game-based learning, problem solving system, scaffolding, systematic computational strategies, aesthetic experience, designed-based learning, embodied learning, HCI teaching, storytelling and universal design for learning.

Game-based learning describes an environment focused on achieving particular learning objectives through gameplay (Kirriemuir and McFarlane, 2004). According to Boyle et al. (2016) and Plass et al. (2015), game-based learning applications are an increasingly important approach in educational interventions due to their ability to keep learners motivated to play and interact with the application or learning environment.

2.3 Escape Rooms

Over the years, educators have implemented game-based learning through educational escape rooms to enhance student motivation and engagement, introduce experiential learning, and divide large tasks into more simple phases (Järveläinen and Paavilainen-Mäntymäki, 2019; Veldkamp et al., 2020; Ross and de Souza-Daw, 2021; Menon et al., 2019). Escape rooms refer to real-life puzzle adventure games where players are "locked" in a room and given puzzles to solve before the time is up (Breakout, 2018; Nicholson, 2015; Clarke et al., 2017). According to Clarke et al. (2017) and Nicholson (2015), escape rooms have the following characteristics: 1) game based on adventure or fantasy, 2) based on theme of escape or rescue, 3) solving puzzles using resources within the game, 4) specific time limit, 5) team based, 6) coooperative rather than competitive, 7) accomplishing a specific goal of escape or rescue, 8) strategising moves that impact game outcomes and 9) learn by doing. Today escape room games can be played in a classroom, online, through board games, or even from a box. Virtual escape rooms refer to games where players attempt to "escape" a virtual environment by solving a mystery or series of puzzles (Coffman-Wolph et al., 2017).

2.4 The Use of Escape Rooms to Facilitate Computational Thinking

Apart from the motivational aspect of escape rooms, it has levels of progression which educators can use to introduce the different components of CT (Menon et al., 2019). Menon et al. (2019) systematically analysed available games with an escape theme meant to develop CT skills. They found three mediums that were used to implement games which include unplugged board games, games with computer programming and games using robotics. Menon et al. (2019) found that a limited number of these games meet the characteristics of an escape room listed by Nicholson (2015) and Clarke et al. (2017). They also found that very little research has been conducted on the impact of virtual escape rooms on the learning of CT. Furthermore, Lathwesen and Belova (2021) highlight the need for practical design frameworks for virtual escape rooms for education. Only one such framework was found and is discussed below.

2.5   The EscapED Framework for Creating Educational Escape Rooms

Clarke et al. (2017) developed a framework for creating educational physical escape rooms, referred to as the escapED framework. Figure 1 below shows a graphical presentation of the framework.

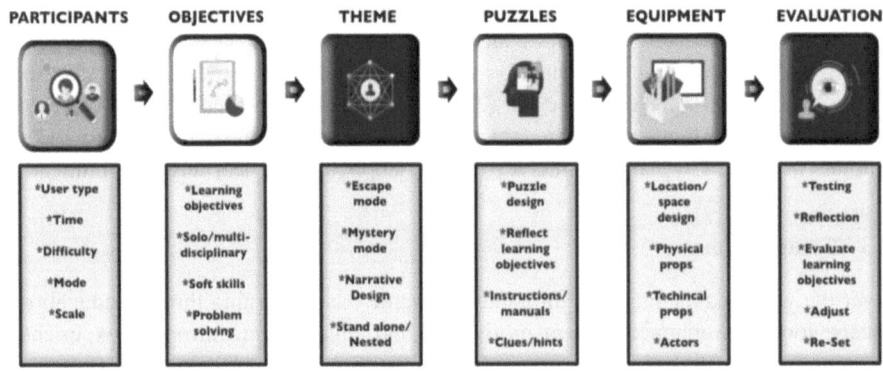

Fig. 1. The EscapED framework (Clarke et al., 2017)

The framework consists of six areas: Participants, Objectives, Theme, Puzzles, Equipment, and Evaluation:

Participants: The first step of the escapED framework involves developers conducting a user assessment, including details such as the target audience. It includes deciding on the length of the escape room experience, the difficulty level, whether the mode will be cooperative or competitive and the number of players it should accommodate.

Objectives: Learning objectives is necessary in order to create a meaningful educational game. The objectives can be worked into various aspects of the game such as the theme, its puzzles and chosen mode to help structure the learning outcomes. It also includes taking decisions such as the discipline or disciplines to design for, the soft skills to be developed (if at all) and developing problem solving challenges to make the game interesting.

Theme: This step involves considering the player's motivations, game story, whether the game is part of a larger, nested experience and content to bring a compelling game experience for the players.

Puzzles: The fourth step of the escapED framework involves developing puzzles and activities with which the players will interact during the game. The puzzles should reflect the the learning objectives. Hints and clues should be used throughout the game as well as instructions or manuals to help guide the learners.

Equipment: The fifth step of the escapED framework involves deciding on the equipment used throughout the game to support the game experience. The physical location should be considered, physical and technical props as well as actors that help concrete the experience as believable.

Evaluation: The sixth and final step of the escapED framework is to consider how the game experience will be evaluated. The game should be tested before implemented, formal evaluation should be used to evaluate if learning took place; adjustment should be considered based on feedback and finally, use a sheet to be checked to ensure everytning is in place before another play-through.

The escapED framework has been implemented in different settings. Löffler et al. (2021) used it to implement an escape room to raise awareness about cybersecurity. Tzima et al. (2021) used the framework to develop a challenging escape game about the local cultural heritage, where players had to solve riddles about the cultural asset of watermills. The escapED framework is used as the basis for the suggested framework developed in this study. The next section explains the methodology followed to develop and evaluate the framework.

3 Methodology

This study followed a Design-Based Research (DBR) strategy. Design-based Research encompasses a range of methodologies that strive to generate new theories, artifacts, and practices related to teaching and learning within authentic environments, while also possessing the potential to be adopted elsewhere (De Villiers and Harpur, 2013). It focuses on an iterative approach that not only evaluates a product or intervention but also strives to enhance the innovation and produce design principles that can steer comparable research and development efforts (Reeves, 2005). There are various DBR models that that were considered, however, for purposes of this study, the focus was on the model proposed by (De Villiers and Harpur, 2013) (see Fig. 2).

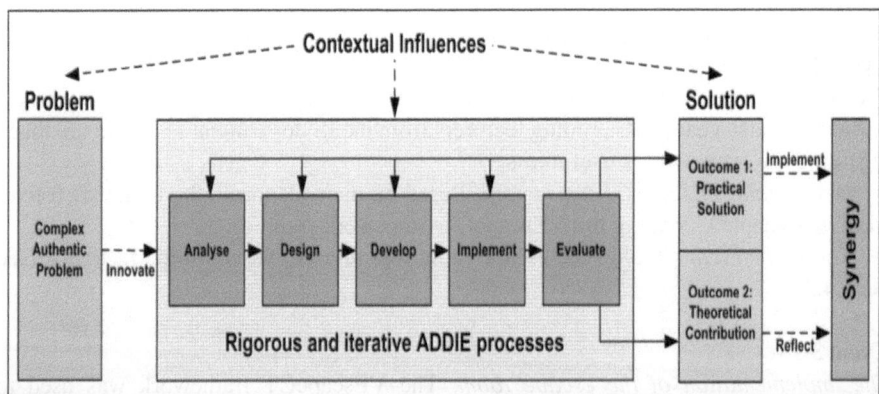

Fig. 2. Generic Model of Design-Based Research Process within a Context (De Villiers and Harpur, 2013)

The model draws inspiration from the well-established ADDIE Design Model encompassing the stages of analysis, design, development, implementation, and evaluation, with a strong emphasis on maintaining rigor. It commences by addressing a complex problem and the necessity for innovation, therefore employing a pragmatic approach to developing the solution while maintaining a synergy between practice and theory and between design and research (De Villiers and Harpur, 2013). The study was conducted at an IS department at a South African University. The first-year programming students and lecturers were participants in this study. Three design cycles were conducted and are described below.

Cycle 1

We used the escapED framework of Clarke et al. (2017) to design and implement the first version of the escape room.

The development and implementation of the escape room. The escape room and implementation of it were done by taking into account the different components of the escapED framework discussed in Sect. 2.5 and adjusting it to a virtual escape room.

The participants. A total of fifteen students participated during the first cycle of the study, of which only four students' data were usable (n = 4), thus resulting in a 26,66% response rate. The activities within the cycle that each participant had to complete included a CT pre-test, playing a virtual escape room, getting a small lecture on how the escape room link to computational thinking and a CT post-test (the same as the pre-test).

The pre-and post-tests. These tests contained 12 multiple-choice questions with four options each, testing theoretical knowledge about the different components of CT (see Sect. 2.1). The responses for each question were categorized according to students with and without INF113 as a module. INF113 is one of the modules offered by the Department of Informatics for undergraduate students, which includes computational thinking in its curriculum. Not all students in the programming course take INF113, and the assumption is that students who did the module, will have a better existing knowledge of CT.

Cycle 2

The participants. To adjust the framework and evaluate the escape room, we interviewed two first-year programming lecturers from the IS department to obtain feedback regarding the framework's usefulness.

Develop the VEscapeCT framework. Based on their feedback, the escapED framework was adapted, creating the VEscapeCT framework (see Fig. 3).

Revise the escape room. The escape room was revised, based on the feedback (see Sect. 4.2).

Cycle 3

The implementation of the escape room. The VEscapeCT framework was used to implement the second version of the escape room.

The participants. A total of 82 students participated during the third cycle of the study, of which 62 students' data were usable (n = 62), thus resulting in a 75,61% response rate. The activities within the cycle were the same as in cycle 1, and the results were again categorised according to students with INF113 and without. In addition, upon completing both the pre- and post-test questionnaires and the virtual escape room, the researcher distributed an evaluation form to determine how the participants experienced the overall process.

Refine the VEscapeCT framework. Final changes were made to the VEscapeCT framework, based on the findings and feedback from students (see Sect. 4.4).

4 Results

4.1 Cycle 1

The Escape Room: The escape room that was developed and implemented in Cycle 1 is presented below. The development was based on the escapED framework of Clarke et al. (2017). The narrative revolved around a hacker who planted a virus on a computer. It consisted of 6 puzzles all related to the components of computational thinking (Table 1).

Table 1. Escape room cycle 1

No	Puzzle Description	CT component
1	Break the code to find the new password	Decomposition
2	Complete the sequence to enable the firewall	Pattern Recognition
3	Identify the additional anti-virus software programs that have been installed due to the virus by identifying 5 unique symbols. Each symbol represents an anti-virus software type	Abstraction
4	Complete the crossword to identify the anti-virus software that needs to remain on the computer	Abstraction
5	Update the anti-virus software by completing the algorithm	Algorithmic Thinking
6	Open your incoming mail but avoid emails from unknown sources	Pattern Recognition

Results: Due to the number of respondents that participated in the first round of the study (n = 4), the researcher could not conduct a t-test. As a result, the researcher calculated an average for each group (see Table 2 for more details).

Table 2. Pre-and Post-Test Analysis of cycle 1

		Average (%)	
Group	# Students	Pre-test	Post-test
Respondents with INF113	2	63	71
Respondents without INF113	2	67	63

The pre-test analysis showed that the group of students without INF 113 achieved a higher average; however, this does not indicate any significance because they might have guessed the answers due to their lack of knowledge. By looking at the post-test analysis, it is clear that the group of students with INF 113 achieved a higher average while improving by 8% compared to the pre-test. Finally, the researcher found it difficult to determine whether the virtual escape room significantly impacted the post-test results due to the number of students participating.

4.2 Cycle 2

To refine and evaluate the framework and escape room, we interviewed two first-year lecturers from the IS department to obtain feedback regarding its usefulness.

The lecturers gave valuable feedback on the framework and the escapED framework was adjusted to create the VEscapeCT framework. Adjustments were made to the s participants, puzzle and equipment components shown in red (see Fig. 3). As technical knowledge is required for a virtual escape room, more time should be spent on defining technical user requirements from the *participants*. The "Difficulty" step was moved from the Participants section to the Puzzles section in the framework. A great escape room is designed with various puzzles, each containing a different difficulty level. The Equipment section has been renamed to Virtual Equipment to reflect its virtual nature.

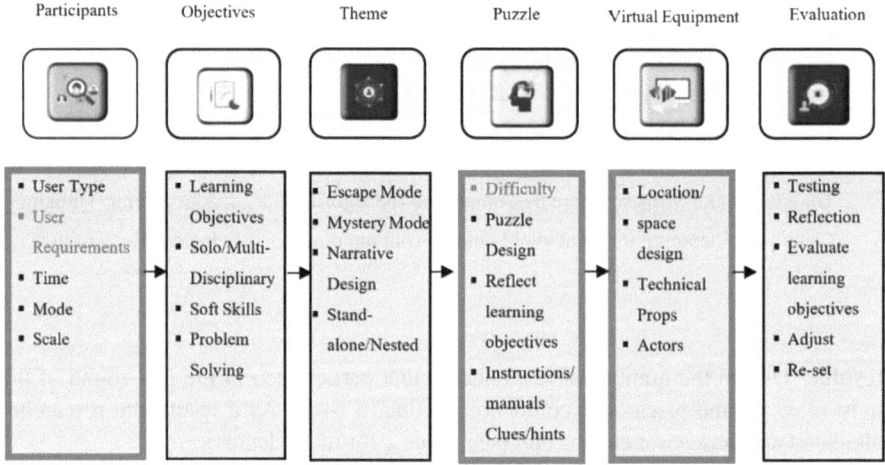

Fig. 3. VEscapeCT Framework (Adapted from Clarke et al. (2017))

Apart from feedback on the framework, the lecturers also commented on the different puzzles in the escape room. For example, they pointed out that puzzle 1 did not illustrate decomposition but rather abstraction. They also mentioned that it is not always possible to link a puzzle with only one CT component. This is in line with computational thinking problem solving approaches, as it always involves most of the components Based on the comments and recommendations, we created a new version of an escape room. The revised version's narrative revolved around a student who planted a virus on his friend's memory stick (Table 3).

The table below provides a few examples of the puzzles (Table 4).

Table 3. Virtual escape room version 2

No	Puzzle Description	CT component
1	Break the code to find the new password	Pattern recognition
2	Unlock the correct permission	Abstraction
3	Run the command line to enable the firewall	Abstraction
4	Update anti-virus	Algorithmic Thinking + Decomposition
5	Provide new product code for expired anti-virus	Pattern Recognition + Decomposition
6	Remove the files copied from the friends flash	Pattern Recognition

Table 4. Examples of the puzzles from the virtual escape room version 2

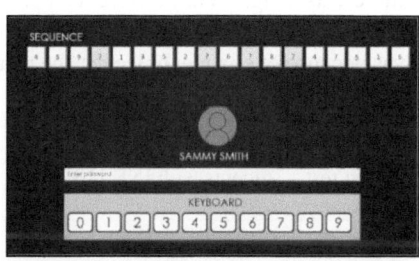

Puzzle 1

Component: Pattern Recognition

Description: The first puzzle was designed to illustrate the concept of pattern recognition. Students were presented with a partially completed pattern which they were instructed to complete. The missing numbers formed the new login details

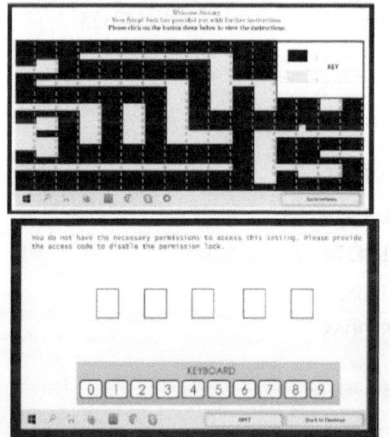

Puzzle 2

Component: Abstraction

Description: The second puzzle was designed to illustrate the concept of abstraction. Students had to use the top screen to find the code necessary to solve the puzzle in the bottom screen

4.3 Cycle 3

The VEscapeCT framework was used to implement the second version of the escape room. Eighty-two students participated during the third cycle of the study, of which only sixty-two students' data were usable ($n = 62$), thus resulting in a 75,61% response rate. For the researcher to determine how each group performed during each test, an average was calculated per test (see Table 5 for more detail):

Table 5. Comparing averages

Group	# Students	Average (%)	
		Pre-test	Post-test
Respondents with INF113	36	82	85
Respondents without INF113	26	77	80

For the researcher to determine whether there was a significant difference between the averages calculated above, a t-test was conducted. The null hypothesis of the test stated that the difference between the means was because of chance; however, the results obtained from the t-test were as follows (see Tables 6 and 7):

Table 6. Results of the t-test of pre and post test – students with INF 113

	Pre-test	Post-test
Mean	0,824074074	0,851851852
Variance	0,024036663	0,019547325
Observations	12	12
Hypothesized Mean Difference	0	
df	22	
t Stat	-0,46091903	
P(T < = t) one-tail	0,324690415	
t Critical one-tail	1,717144374	
P(T < = t) two-tail	0,64938083	
t Critical two-tail	2,073873068	
Confidence Level	95	
Significance Level	0.05	

At a 0.05 percent level of significance and degree of freedom (df, 22), the p-value of 0,64938083 is more than the significance level of 0.05. Therefore, there is no significant difference between the means obtained from the two tests, as a result, the researcher accepts the null hypothesis. The researcher confirms that the difference between the averages obtained from each group occurred by chance

Table 7. Results of the t-test for the pre and post-test - students without INF 113

	Pre-test	Post-test
Mean	0,769230769	0,804487179
Variance	0,036578806	0,028633226
Observations	12	12
Hypothesized Mean Difference	0	
df	22	
t Stat	-0,47826087	
P(T < = t) one-tail	0,318591451	
t Critical one-tail	1,717144374	
P(T < = t) two-tail	0,637182903	
t Critical two-tail	2,073873068	
Confidence Level	95	
Significance Level	0.05	

At a 0.05 percent level of significance and degree of freedom (df, 22), the p-value of 0,637182903 is more than the significance level of 0.05. Therefore, there is no significant difference between the means obtained from the two tests and, as a result, the researcher accepts the null hypothesis. The researcher confirms that the difference between the averages obtained from each group occurred by chance

Although there was no significant difference between the averages, and we concluded that the averages occurred by chance, both groups achieved a higher average in the post-test questionnaire compared to the pre-test questionnaire.

Upon completing both the pre- and post-test questionnaires and the virtual escape room, we distributed an evaluation form to determine how the respondents experienced the overall process.

The feedback that was obtained was very positive. The main themes that emerged from these responses are that of enjoyment and fun, finding the escape room a good challenge and that it is conducive for learning (see Table 8 below):

4.4 The Final VEscapeCT Framework

After cycle 3 the framework was further refined, based on the observations and feedback from students. The component *Equipment* was removed to make place for *Platform*, which includes the aspects of the virtual environment which should be considered: cost of the software, accessibility of the platform in terms of availability and user-friendliness, ease-of-use of the platform, and the functions available that can be used to create an escape room. Evaluation now includes both testing and implementation. The escape room should be tested before and after implementation, reflected on and adjusted if needed. These components are illustrated in Fig. 4 below.

Although the results did not provide conclusive evidence that CT learning occurred, students showed evident enjoyment and a willingness to learn more about CT after their

Table 8. Feedback from students on the escape room experience

Remark	Theme
"The escape room was pretty fun to do:-)"; *"Great opening to Semester 2"*; *"The escape room was fun and interesting"*; *"I loved the escape room even though I'd already learnt about computational thinking from INF113"*; *"I really enjoyed the exercise, I wish similar exercises can be made that are interesting like this one."*	Enjoyment, fun
"The escape room activity was interesting, but I do not think it aided to my understanding of computational thinking. It was just an interesting exercise that required a lot of thinking. I do believe I already understood what computational thinking was."; *"I really enjoyed this practical. It was very interesting and I learnt quite a bit."*; *"I loved the interactive experience that the escape room provided! It was so much fun and really enabled me to put the theory behind computational thinking into practice. I would definitely participate in a similar event again."* *"I enjoyed the escape room, it was fun even though it was a bit challenging* *"It was a really fun and creative way to go about teaching a concept which ordinarily is thought to be very linear."*	Conducive for learning

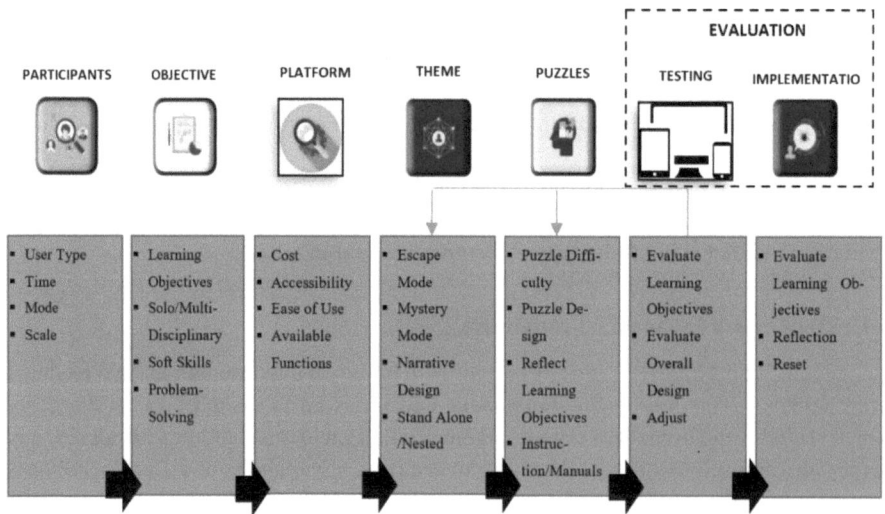

Fig. 4. Final VEscapeCT Framework

escape room experience. Virtual escape rooms can be used by educators to motivate students or to get students more interested in a particular subject. If educators wish to create the ultimate learning experience, researchers recommend implementing virtual escape rooms alongside other educational interventions. Such a learning environment can provide students with the necessary motivation and a comprehensive understanding of CT concepts (Sánchez-Ruiz et al., 2022).

This study contributed towards the body of knowledge on the creation of virtual escape rooms in three ways: 1) It provides an example of an escape room design and how it links to CT components, 2) it provides guidelines on how to design, implement and evaluate CT virtual escape rooms in the form of the VEscapeCT framework, 3) It illustrates how the impact of the implementation can be tested using a CT assessment test. The CT test used in the pre-and post-test focuses more on a high-level understanding of the different components of CT and less on the application thereof. It will be interesting to do another implementation of the escape room and use an existing CT competency test (e.g. the test developed by Lai (2021)).

5   Conclusion

This study suggests a framework for designing, implementing and evaluating an escape room to teach CT in Higher Education. The framework was developed following the DBR steps, using three design cycles in the context of a first-year IS programming student group. Feedback from students shows evident enjoyment and motivation to learn more about CT.

This study has a few limitations. Due to COVID, we could not monitor students as they participated in the escape room and therefore had little control over the level of involvement in the intervention. For this reason, only four of the 15 students who started the activities, completed all the activities, and could be included in the study. For the same reason, in cycle 3, the data of only 62 of the 82 students could be used.

Finally, future research includes further refining the framework with more detailed CT guidelines. Also, researchers can use this framework, as it is generic enough, to build subject-specific virtual escape room frameworks for other subjects, especially those with low student motivation. The value of an escape room lies in its potential to engage students, challenge them and provide a fun way of learning.

References

Angeli, C., et al.: A K-6 computational thinking curriculum framework: implications for teacher knowledge. J. Educ. Technol. Soc. 19(3), 47–57 (2016)

Akour, M., Alsghaier, H., Aldiabat, S.: Game-based learning approach to improve self-learning motivated students. Int. J. Technol. Enhanced Learn. 12(2), 146–160 (2020)

Barr, V., Stephenson, C.: Bringing computational thinking to K-12: what is Involved and what is the role of the computer science education community? ACM Inroads 2(1), 48–54 (2011)

Berland, M., Lee, V.: Collaboratibe strategic board games as a site for distributed computational thinking. Int. J. Game-Based Learn. 1(2), 65–81 (2011)

Boyle, E.A., et al.: An update to the systematic literature review of empirical evidence of the impacts and outcomes of computer games and serious games. Computers and Education 5(2) (2016)

Breakout: History of Escape Rooms (2018). [Online] Available at: https://breakoutgames.com/escape-rooms/history. Accessed 30 August 2021

Clarke, S., Peel, D., Arnab, S., Morini, L., Keegan, H., Wood, O.: EscapED: a framework for creating educational escape rooms and interactive games for higher/further education. Int. J. Serious Games 4(3), 73–86 (2017)

Coffman-Wolph, S., Gray, K.M., Pool, M.A.: Design of a Virtual Escape Room for K-12 Supplemental Coursework and Problem Solving Skill Development. West Virginia, ASEE Zone II Conference (2017)

Connolly, T., Hainey, T., Stansfield, S.: Development of general framework for evaluating game-based learning. s.l. In: Proceedings of the European Conference on Game-Based Learning (2008)

De Villiers, R., Harpur, P.: Design-based research – the educational technology variant of design research: Illustrated by the design of an m-learning environment. s.l., ACM (2013)

Ding, D., Guan, C., Yu, Y.: Game-based learning in tertiary education: a new learning experience for the generation Z. Int. J. Info. Edu. Technol. 7(3), 148–152 (2017)

Durak, H., Yildiz, Y., Fatma, G.K., Yilmaz, R.: Examining the relationship between digital game preferences and computational thinking skills. Contremporary Edu. Technnol. 8(3), 359–369 (2017)

Hsu, T.-C., Chang, S.-C., Hung, Y.-T.: How to learn and how to teach computational thinking: Suggestions based on a review of the literature. Comput. Educ. 126(2018), 296–310 (2018)

Järveläinen, J., Paavilainen-Mäntymäki, E.: Escape Room as Game-Based Learning Process: Causation - Effectuation Perspective. s.l., HICSS (2019)

Kazimoglu, C., Kiernan, M., Bacon, L., Mackinnon, L.: A serious game for developing computational thinking and learning introductory computer programming. Soc. Behav. Sci. 47, 1991–1999 (2012)

Kirriemuir, J., McFarlane, C.A.: Literature Review in Games and Learning. FutureLab, Bristol, UK (2004)

Lamprou, A., Repenning, A.: Teaching how to teach Computational Thinking. ITiCSE, 69–74 (2018)

Lai, R.P.: Beyond programming: A computer-based assessment of computational thinking competency. ACM Trans. Comp. Edu. (TOCE) 22(2), 1–27 (2021)

Lathwesen, C., Belova, N.: Escape Rooms in STEM teaching and learning—prospective field or declining trend? a literature review. Education Sciences 11(6), 308 (2021)

Löffler, E., Schneider, B., Zanwar, T., Asprion, P.M.: Cysecescape 2.0—a virtual escape room to raise cybersecurity awareness. Int. J. Serious Games 8(1), 59–70 (2021)

Menon, D., Romero, M., Vieville, T.: Computational thinking development and assessment through tabletop escape games. Int. J. Serious Games 6(4), 3–18 (2019)

Mohaghegh, M., McCauley, M.: Computational Thinking: The Skill Set of the 21st Century. Int. J. Comp. Sci. Info. Technol. 7(3), 1524–1530 (2016)

Nicholson, S.: Peeking Behind the Locked Door: A Survey of Escape Room Facilities, s.l.: s.n (2015)

Nicholson, S.: Creating engaging escape rooms in the classroom. Child. Educ. 94(1), 44–49 (2018)

Plass, J., Homer, B., Charles, K.: Foundations of game-based learning. Educational Psychologist 50(4), 258–283 (2015)

Reeves, T.C.: Design-based research in educational technology: Progress made, challenges remain. Educ. Technol. 45(1), 48–52 (2005)

Ross, R., de Souza-Daw, A.: Educational escape rooms as an active learning tool for teaching telecommunications engineering. Telecom 2, 155–166 (2021)

Sánchez-Ruiz, L.M., López-Alfonso, S., Moll-López, S., Moraño-Fernández, J.A., Vega-Fleitas, E.: Educational digital escape rooms footprint on students' feelings: a case study within aerospace engineering. Information 13, 478 (2022)

Selby, C., Woollard, J.: Computational thinking: the developing definition. University of Southampton (2013)

Tatar, C., Eseryel, D.: A literature review: Fostering computational thinking through game-based learning in K-12. In: The 42nd Annual Convention of The Association for the Educational Communications and Technology, pp. 288–297 (2019)

Veldkamp, A., van de Grint, L., Knippels, M.-C., van Joolingen, W.: Escape education: A systematic review on escape rooms in education. Educ. Res. Rev. **31**(2020), 18 (2020)

Wing, J.: Computational Thinking. Commun. ACM **49**(3), 33–35 (2006)

Wing, J.M.: Computational thinking and thinking about computing. Philosoph. Trans. Royal Soc. London a: Mathemat. Physi. Eng. Sci. **366**(1881), 3717–3725 (2008)

Wing, J.M.: Research Notebook: Computational thinking -what and why?, s.l.: The Link Magazine (2011)

Zaibon, S.B., Shiratuddin, N.: Mobile game-based learning (mGBL) engineering model as a systematic development approach. In: Global Learn, pp. 1862–1871. Association for the Advancement of Computing in Education (AACE) (2010, May)

Value of Explicit Instruction in Teaching Computer Programming to Post-graduate Students: The Kirkpatrick Training Evaluation Model

Pakiso J. Khomokhoana(✉) and Ruth Wario

Department of Computer Science and Informatics, University of the Free State, Bloemfontein, South Africa
khomokhoanap@ufs.ac.za

Abstract. The effectiveness of ways of teaching in various disciplines including Computer Science (CS) programming continues to be a debatable issue. In this study, we followed the Kirkpatrick training evaluation model to better understand how Explicit Instruction (EI) interventions in teaching computer programming to post-graduate CS students can be evaluated. This understanding can be fundamental in informing how improvements can be made in the future teaching of computer programming to create a better sense of quality and robustness. The aim of this study was threefold. Firstly, to provide an overview of the Kirkpatrick model. Secondly, to relate the four levels of this model to the teaching of computer programming. Thirdly, to evaluate the use of EI interventions as a teaching approach through the model. This study followed an integrated methodological approach where data was collected through asking questions (individual semi-structured interviews). Thematic analysis of the collected data revealed that our participants found a lot of value from the EI interventions, and hence appreciate them. Overall, the findings of this study are consistent with those of many researchers that EI is one of the effective teaching and learning strategies, not only in other disciplines, but in computer programming as well.

Keywords: Computer programming · Explicit instruction · Explicit instruction interventions · Teaching strategies · Evaluation · Computer science education · The Kirkpatrick Training Evaluation Model

1 Introduction

Instructors in various disciplines including Computer Science (CS) programming continue to debate on the effectiveness of ways of teaching. As such, it is necessary to continuously evaluate strategies that instructors use in their teaching [11, 12]. This can help the trainer (referred to as "instructor" hereafter) to know whether the training that he/she has conducted has been effective or not [3, 10]. Furthermore, the evaluation can help to ensure that the teaching is meeting students' learning needs, and to recognise areas where the teaching can be improved [36].

H. E. Van Rensburg et al. (Eds.): SACLA 2023, CCIS 1862, pp. 18–33, 2024.
https://doi.org/10.1007/978-3-031-48536-7_2

Though originally developed for the business discipline, one popular evaluation model used within higher education [45] was designed and developed by Donald Kirkpatrick, and presented in his seminal articles published in 1959. This model has been applied within higher education in various disciplines such as the in-service vocational teacher training program [3], leadership training programs for sustainable impact [37], and blended learning environments [17]. The model provides a pragmatic approach for any typically complex evaluation process [5]. The Kirkpatrick model consists of four levels as follows [27, 28]:

- Reaction (Level 1) – How did the students perceive the training?
- Learning (Level 2) – Were the students able to learn from the training?
- Behavior (Level 3) – Were the participants able to transfer what they learned during the training?
- Results (Level 4) – What are the tangible results from the training?

According to Scherer et al. [49], computer programming is recognized as a universally significant discipline. However, the systematic evaluation of the effectiveness of instructional approaches and conditions that enhance the acquisition of programming knowledge and skills has received little attention [20, 33]. As such, it may be necessary to evaluate some of the approaches used in teaching computer programming. One of these approaches is Explicit Instruction (EI) – pioneered by Rosenshine and Stevens [46]. This type of instruction is defined as an approach that can help students to better learn and understand a variety of concepts in numerous disciplines [21]. The Computer Programming discipline is not an exception. The effectiveness of EI has already been confirmed many times [25, 44]. The underlying principle of EI is that the transfer of knowledge is conducted in a structured, systematic and planned way where concepts are treated in the order from simple to complex, and from easy to difficult. Any teaching strategy employing EI follows steps that are sequenced and strongly integrated. Literature suggests that in using EI, it is equally important to clarify the learning objectives and intended outcomes, identify key ideas, and determine students' prior knowledge [21]. Greene [19] summarises the steps of EI as shown in Table 1.

All activities performed as part of EI are aimed at creating some form of cognitive scaffolds for the student [2], and to reduce working memory overload [25]. According to Bocquillon et al. [6, p.12], students taught using this approach can learn "without conscious effort" and "without taking up the memory working space". Tshukudu and Jensen [53] found that EI interventions deepen understanding of programming concepts, hence concluded that these interventions are effective. Irrespective of its effectiveness, this approach is also associated with limitations. These include; encouraging students to sit passively and engage in rote learning [22], encouraging students to over-rely on memorisation, encouraging students to lose interest (boredom) and limiting students' creativity [9, 26].

Considering the limitations associated with EI interventions, this study attempts to answer the research question: *How can the Kirkpatrick training evaluation model be used to evaluate the teaching and learning of a post-graduate computer programming module?* To fully answer this question, it was broken down into two subsidiary questions as follows:

Table 1. A Summary of the EI Steps

Step	Name	Description
1	Identify clear and specific objective(s)	Instructors think and decide on what they want to teach, and hence determine the definite objectives to achieve in the teaching subject in question
2	Break the information into chunks	Instructors break down concepts into small and meaningful segments that students will be able to easily grasp and understand without having to engage a lot of cognitive load
3	Model with clear explanations	Instructors explain or demonstrate a skill that students should acquire in the same way that students will practice it. This is normally achieved through using considered language and being as simple and natural as possible
4	Verbalize the thinking process	Instructors perform a think-aloud of what is going on in their minds as they model explanations made and skills to be fostered to students
5	Provide opportunities to practice	Instructors allow students plenty of time to practice skills imparted to them
6	Give feedback	Instructors give their comments (affirmative or constructive) on how students performed on the given activity or a set of activities

- What is the Kirkpatrick model of training evaluation?
- How can using EI as an approach, be evaluated through the Kirkpatrick model of training evaluation? [Integrating the elements of EI with the model].

In the remainder of this paper, a theoretical framework guiding this study together with a review of relevant background literature are presented in Sect. 2. Section 3 presents the pedagogical activities carried out throughout the EI interventions. This is followed by a discussion of the research design and methods in Sect. 4, and a presentation and interpretation of the results in Sect. 5. Conclusions and recommendations for future research are presented in Sect. 6.

2   Theoretical Framework

Kirkpatrick and Kirkpatrick's [27] training evaluation model provides the theoretical framework for this study. In the following sub-sections, the four basic levels of training evaluation of Kirkpatrick and Kirkpatrick's model, together with examples of how these steps are typically executed by computer programmers, are discussed in more detail. It is also important to note that answers to all the evaluation questions under each phase of the model can be ascertained through using various data collection techniques such as tests, quizzes, surveys, questionnaires, observations, expert or peer review, interviews and focus groups [39, 52].

2.1 Reaction

In every kind of training, it is key to work in close cooperation with either students, participants or trainees (referred to as "students" hereafter) in order to gauge the level of satisfaction or dissatisfaction that they might be experiencing along the way. Kirkpatrick and Kirkpatrick [27] provide several questions that can be asked to establish the reaction of students throughout the training: Did the students like and enjoy the training? Did they consider the training relevant? Was the training a good use of their time? Did students like the training style? How engaged were students during the training? How much contribution did students make during the training, and how actively did they do that? How was the overall reaction of students to the training? How well did the participants perceive the training? How satisfied are the students with feedback received as a result of the training? How satisfied are students with communication of training's objectives? How satisfied are students with quality of the overall training?

Considering the reaction of students to pair programming as an intervention in teaching computer programming, Salleh et al. [47] report that students get more satisfaction when using pair programming than when they work individually. Furthermore, if students are satisfied by the computer programming intervention(s), they may develop a positive attitude towards the subject whose content was covered during the training. As an example, in exploring the relationship among students' attitudes toward programming, gender and academic achievement in programming, Baser's [4] correlations revealed a positive relationship between students' achievements and their attitudes towards computer programming.

2.2 Learning

Irrespective of the type of training engaged in, it is expected that students will learn something from such training. This learning typically increases the skill level, capability, commitment, confidence, attitude and knowledge of the student in question from before, to after the learning experience. Kirkpatrick and Kirkpatrick [29, p. 6] define these constructs as follows:

- *Skill* – the degree to which students know how to do something or perform a certain task, as illustrated by the phrase, "I can do it right now".
- *Knowledge* – the degree to which participants know certain information, as characterized by the phrase, "I know it".
- *Attitude* – the degree to which training participants believe that it will be worthwhile to implement what is learned during training on the job. Attitude is characterized by the phrase, "I believe it will be worthwhile" to do this in my work.
- *Confidence* – the degree to which training participants think they will be able to do what they learned during training on the job, as characterized by the phrase, "I think I can do it on the job".

- *Commitment* – the degree to which learners intend to apply the knowledge and skills learned during training to their jobs. It is characterized by the phrase, "I will do it on the job".

According to Kirkpatrick and Kirkpatrick [27], to effectively assess this increase, trainers can test the knowledge prior and post training. The Kirkpatricks' further provide several questions that can be used to assess whether the learning occurred during the training: Did the students learn what was intended to be taught? Did the students experience what was intended for them to experience? What is the extent of advancement or change in the students after the training, in the direction or area that was intended? Did my knowledge/skills increase because of the training? Has using the new knowledge and skills helped me to improve my work?

During the learning process in computer programming, Martin and Shafer [35] recommend that the code should be modularized. This entails ensuring that the code is being structured into small parts which minimize gross coupling and simplify understanding. In a study comparing learners' knowledge, behaviors, and attitudes between two instructional modes of computer programming, Sun et al. [50] identified listening to the instructor, followed by either the behavior of discussing with peer or taking notes as some of the most effective ways that students use to learn. According to Wilson [54], and for programming students, the stated characteristics are linked to student comfort and confidence. In this regard, Alturki [1] further argues that programming students who are comfortable and confident ultimately perform well overall. However, it cannot always be guaranteed that when students engage in these practices, they are learning. As part of learning in computer programming, researchers suggest that students should engage with both the theoretical and the practical aspects of the concepts taught [8, 32, 51].

2.3 Behaviour

As a result of the learning experiences an individual goes through, his/her behaviour is expected to change to indicate how well he/she has learned from the training. However, individuals who measure behaviour should be able to tolerate delays because this type of measurement is a long process [10]. Kirkpatrick and Kirkpatrick [27] provide questions such as the following to ascertain whether the behaviour has changed after the training:

- Did the students put their learning into effect when back on the job?
- Were the relevant skills and knowledge used?
- Was there noticeable and measurable change in the activity and performance of the students when back in their roles?
- Was the change in behavior and new level of knowledge sustained?
- Would the student be able to transfer their learning to another person?
- Are the students aware of their change in behavior, knowledge, and skill level?

As part of behaviour exhibition, Sun et al. [50] identified behaviours such as recall of programming knowledge, feeling of learning experiences, and attitudes towards programming. These were revealed after engaging students in unplugged programming activities. In unplugged computer programming, one of the other behaviours that students exhibit when working on the activities is to be engaged, and this engagement enhances students' satisfaction and enjoyment [18]. As part of behaviour exhibited during and after the teaching intervention, students can consolidate their knowledge, problem solving skills and logical reasoning [7, 48]. Moreover, as computer programming cultivates higher order thinking abilities, students could exhibit higher computational thinking and logical thinking behaviours after the training [41].

2.4   Results

Following from the learning experiences and the change of behaviour of a student during the training process, he/she is expected to also show results (positive or negative) in relation to the training that he/she underwent. This, in turn, gives an indication of the effectiveness of the training. According to [15], level 4 is more involving (difficult and time-consuming) compared to the other levels. However, it provides valuable information that is associated to the worth of a learning and performance process. Kirkpatrick and Kirkpatrick [27] provide questions such as the following to indicate whether the results of training are positive or not:

- Is the student still happy after the training?
- What was the impact of the training on his/her performance?
- Are there any benefits that he/she gained from the training?
- To what degree did the targeted outcomes occur because of the training event and subsequent reinforcement?

In reviewing the effectiveness of interventions made in computer programming courses, several studies use pass rates as a measure of the effectiveness [1, 24]. As a result of the teaching intervention in computer programming, students may be seen to prefer other teaching and learning styles (dynamic, individualistic, peer, etc.) than others that may have been used during the intervention [13, 55]. On reviewing the effects of pair programming as an intervention in CS education, Hanks et al. [23] reported increased pass and retention rates, enhanced student confidence in solutions, and improvement in learning outcomes as some of the benefits of the intervention. In evaluating the impact of various forms of content interventions (including the choice of programming language) for teaching object-oriented programming on students' learning of fundamental and object-oriented programming concepts, Kunkle and Allen [31] report that students showed an overall good performance as a result.

3   Pedagogical Intervention

There are various pedagogical activities that the first author carried out from inception throughout the intervention. Initially, he investigated all the instructional goals of the module for which he made this intervention. However, due to time limitations, he focused on the main goal (ability to design, code and implement mobile applications) where three concepts namely: animation, Google maps, and SQLite Database were discussed. For the animation concept, all content was presented to students in a traditional face-to-face class using lecture slides and making the necessary demonstrations. For both the Google maps and SQLite Database concepts, audio-visual lectures were shared with students well before class and summaries of the discussion of these concepts were presented to students in class and they asked questions where they did not understand. In discussing these three concepts, all the explicit steps necessary to achieve the learning goals were elaborated on and demonstrated to students.

To prepare the learning environment and preliminary learning resources, he ensured that all the necessary resources that students would need were available to them. For instance, installation of Android Studio and the necessary software packages on computers in the venue that students used. He also tested this installation before students could start using it to ensure that they would not experience problems when using the software. A module guide detailing all the meeting times and venues was also made available to students upfront, and all content in the module guide was discussed with the students.

He also created an instructional strategy that helped him to achieve the selected instructional goal. This involved creating various learning components such as preparing further lecture notes, designing instructional activities that students completed, documenting strategies to use to ensure that students got engaged in the learning content presented to them (i.e., allow students to practice a task and provide them with timely and descriptive feedback); as well as how students were assessed on the skills acquired. In the main, he kept motivating students by praising and valuing outstanding solutions when giving feedback.

He designed two assessments that tested whether students were able to master the instructional goal specified earlier. In preparing the assessments, he was careful to ensure that they specifically assessed the design, coding and implementation skills linked to various features of the three concepts specified earlier. He also ensured that questions included in the assessments were clear and worded in correct punctuation and grammar. The lectures were presented in a computer lab where students had access to computers installed with the software.

4   Research Design and Methods

4.1 Design

Within the scope of an EI-based research design, we followed an approach based on Plowright's Frameworks for an Integrated Methodology (FraIM) [42]. FraIM advocates that researchers must not necessarily take a philosophical position prior to the beginning of the study. Such a position can be taken as the study develops or even during the interpretation of results. As such, we focused on the collection of narrative data through asking questions. The study population comprised postgraduate Honours CS students (cited as postgraduate students in this paper) from a South African university. The study sample consisted of 14 students registered for an Honours Advanced Programming module. The selection of this sample was purposeful because the first author wanted to improve the teaching and learning strategies for the current and upcoming students for the module. The sample was also convenient since he had easy access to the participants [40]. For the individual interviews at the end of the semester, students were invited to participate during their leisure time as this was not completed during the regular class sessions.

4.2 Data Collection

According to Plowright [42], a data collection strategy using individual interviews can be considered as a way of 'asking questions'. For the individual interviews, all the students were invited. However, only eight of them agreed to participate by completing an informed consent form before the interviews. In these semi-structured interviews, students were asked to share their experiences with the EI interventions that were implemented in the classes for the selected module. Probing questions were asked where necessary. All the proceedings of the interviews were audio recorded after each participant granted us permission to do so. The interview sessions were each scheduled for 60 min.

4.3 Data Analysis

In transcribing and analysing the audio recordings from the interviews, we followed Creswell and Creswell's [16] approach. After transcribing the data, we cleaned it through identifying faults and correcting them accordingly [14]. As the questions were open-ended, the transcripts had several statements that were illogical and repeated. As such, we opted to apply fuzzy validation rather than strict validation. The latter requires the complete removal of invalid or undesired responses) [38]. Fuzzy validation advocates that researchers can correct some data if there is a rationally close match to a known correct answer. After cleaning the data, we acquainted ourselves with it [34] by listening and re-listening to the audio records several times and by thoroughly and repeatedly reading the transcripts. From this acquaintance, we were able to formulate a coding plan in which the analysis would be guided by the data related to our research questions. At this stage, we imported the eight validated transcripts into the Nvivo tool. We then developed codes for each element belonging to each level of the Kirkpatrick's model in the data. To code effectively, researchers can use units of analysis which may include words, sentences, or paragraphs [30]. As such, we coded the data through marking text (e.g., highlighting, underlining, circling, and annotating) within the confines of the stated units of analysis. We then populated our created codes through linking them with the complementary segments of text. In this refinement process, we constantly revised the names of the codes until suitable themes emerged. We considered the frequency of occurrence for each emerging theme.

5 Results and Interpretation

The discussion in this section centres around the elements of the Kirkpatrick model that were identified from the data collected during the implementation of the EI interventions in teaching computer programming to post-graduate students. The discussion is grouped according to the four levels of the model (see Sect. 2).

5.1 Reaction

Analysis of the data collected through the individual interviews uncovered that the participants expressed positive reactions to the EI interventions that were implemented throughout the semester. Eight participants, with a total of 42 occurrences, were satisfied by the interventions.

As evidence that P1 enjoyed the EI interventions, he said: *"Well, what was more interesting at most was, obviously, getting the step-by-step instructions because you could know what you did wrong and what you didn't do wrong"*. In a sense, he was referring to the benefits that were introduced by the EI interventions. This probably might have helped him to create a positive attitude [4] toward the module as in the end he performed well (75% as a final mark for the module). Furthermore, on overall, he expressed to have liked how the interventions were carried out as he remarked: *"In terms of the way everything transitioned, I feel like everything just went well... I wouldn't say we needed an improvement in a way because we gained enough basics with the explicit instructions from the get-go"*. P1 also made a key observation that there should be a time when instructors must transition from being explicit to being implicit: *"The challenge will always be, when do you say is the time for you to start introducing implicit concepts or instruction? ... so, in this case, I feel like the timing was perfect, because we had learned all the necessary skills for us to build onto the understanding of how things work"*. Although the instructor must strike a balance between using explicit and implicit steps in delivering the learning content to students, he/she must also ensure that some students are not bored by the explicit steps [46] if they have already mastered concept(s) under discussion.

Typically, the reaction of P4 to EI Step 6 (give feedback) intervention revealed three key aspects namely; feedback decency, peer learning [43] and social networking. Regarding feedback decency, P4 said: *"It was the first time this year that feedback was given in class. Usually, we just receive an email to say, 'hey guys, your results are out' and that's it ... it was a great strategy"*. This was an overwhelming revelation because generally most of the instructors seem not to treat student feedback with the decency that it deserves. Peer learning [43] and networking, which are the promising methods of effective learning also surfaced in the way feedback was given to students. P4 remarked: *"So even now **I know [Student X] and [Student Y] because of you mentioning them in class**, saying that [Student X] presented this so well ... I was impressed about that ... I started asking him questions"*. This feedback-triggered relationship could last forever among the students.

5.2 Learning

A total of 33 occurrences from the eight participants were identified, where students expressed to have learned from the EI interventions. To indicate that P1 learned from the EI interventions, he remarked: *"After following the steps, I had to perhaps add one extra feature on to the things, given the fact that I had the step by step for all the other features, and I had to draw intuition from all those in terms of how you implement things in the environment, it was the most interesting part because it showed also, that even though I was following the step by step in some way, I was also learning how to implement the things in a specific environment"*. Adding features and drawing on intuition are elements that indicate that P1 engaged with the learning content, and hence learned along the way. The participants further indicated to have learned from the oral presentations that students made in class as part of the EI interventions. P1 remarked: *"Everyone came out of that class understanding how the Google Maps class works*"; while P5 stated: "*I was able to understand Google Maps better from the oral presentations*". P1 also indicated that preparation for the presentations engaged students to learn as well: "*When we were doing the presentation, even before the presentation took place, we had already learned how to implement those concepts*". As evidence that P2 learned, he said: "*Yes, there were examples and they did help me to learn … with the Google Maps, I did learn a lot more, with the presentations as well, you know, just seeing other people doing things and what else is out there*".

When working with the list view and recycler view, P3 was able to identify and learn about deprecated functions: "*There was another example about the list view and the recycler view, and the recycler view was deprecated. So, we had to figure out, how do we apply a list view for it to be like a recycler view?*". Due to the high rate at which the Google team releases Android versions, most of the time the latest versions or Application Programming Interfaces (APIs) are not compatible with the old classes or methods (a concept called deprecation). When designing the User Interface (UI) for the Calculator App, P3 further indicated the elements of learning: "*So we had to do it with another layout. So, I did it with a table layout. So that was interesting to learn that the table layout can also be used to do this*". Participants also indicated that they learned during the time feedback was given to them. P3 remarked: "*You see, okay, this is what I did wrong. And this is how I can do this, how it's supposed to be done, or there are some ways that it could be done, but it's just not this way that you did it*". On giving feedback, the instructor also publicly made references to students who did well on certain concepts, and this helped other classmates to contact them. P6 remarked: "*When I did have a problem with the Google Maps assignment, I did go to one of the students who did well, and then ask him about it, he managed to do it perfectly. So, I learned more from him*".

P4 expressed to have learned in an instance where the zoom concept of Google maps was discussed: "*When you were putting markers on the map, you showed us an example that you can put a marker here, and you can change the zoom value, right, and maybe add 5 or 10, depending on the size that you want to see. So, when you did that, I understood, ouh, so if I have a situation where I'm working with street level view, this is the value I will use. So, it helped me in that sense that I was able to understand and was able to apply it in different scenarios*". In one assignment, students were given a task to read on various functionalities of Google Maps and implement only four of those. P4 specifically indicated that this type of questioning did not only help him to be creative, but also kept him engaged with the learning content as well. He said: "*I don't have so many modules where I'm given that kind of leeway, where I can be creative, and then decide what I would like to do … so it kept me really engaged*". This implies that students engage and hence learn more when they work on learning activities that they choose for themselves. P7 indicated elements of learning by indicating the resources that he engages with when he is faced with some tasks to complete: "*How I go about solving problems like, I'll go to Stack Overflow and GitHub, and then I can clearly look at this and then translate it into this specific use case that I'm being asked for … and I try to make sense of the code and then, from the code, I can clean up a few things …then I can piece everything together regarding that thing*".

5.3 Behaviour

There were various behavioural aspects that were identified in the data. Fourteen occurrences from eight participants indicated various behaviours. The participants indicated a behaviour that is normally seen with experts in the field. They normally consult other sources to get an understanding of concept(s) or answers to questions that might be challenging at the moment. P1 remarked: *"When a concept is not explained explicitly, in a way, I'm obviously forced to go out in the world and see how someone else might have implemented it"*. P2 said: *"There was another student who implemented the search functionality [on Google maps], search something and it appears on a map with buildings and stuff. So that was really interesting. Okay, is this possible? This forced me to go and study what was happening behind the application and how I can develop that application without him"*. P3 remarked: *"And I also realize that there are some things that I didn't actually understand in my application, and how they work. I just searched on the Internet; how do I do this?"*.

Following from the challenging questions that came during the oral presentations, P3 said: *"So when somebody asked the question, and I couldn't come up with the answer, I just saw, okay, there's a gap here. I need to go and do this again"*. Participants also expressed some behaviour that indicates that students are hesitant to be exposed to learning activities that may ultimately help them to effectively learn. For example, oral presentations seem to be effective, but students are not comfortable with them. P2 remarked: *"I wouldn't like presenting. I don't think a lot of people like presenting, but at the end of the day, the results you get from it are worthwhile"*. P3 remarked: *"There are some students, like me, who are not comfortable with presenting. But at the same time, it also forces them to just get out of their shell and just present. It's not always that they're going to be just writing the code [and submit], and nobody's going to ask them any questions"*.

Some participants indicated a typical behaviour that is existent with some students that they would rather try to search for answers by themselves without either asking instructors or other colleagues. P7 remarked: *"This would be determined by how close I am to the person ... I might not be comfortable with him to go and ask; can you show me this? ... In my experience, in most cases, the concepts that I encounter in programming, it's something that I can Google ... I'm averse to going and asking for help ... I've never had a lot of opportunities in programming for me to do that. Because in most cases, there's always something or some resource online that I can go check ... I usually open many tabs, like if I'm googling something, sixteen or eight tabs open"*. It can be seen in this extract that P7 would also consult with a lot of sources at one time in trying to understand a concept or to solve a problem.

5.4 Results

In agreement with Clark [15], this level proved to be more difficult and time-consuming compared to the other levels. What even makes it more difficult is the fact that instructors have little time to interact with the students after the training. This is the case as students prepare for examinations immediately after the training. After the examinations, it is rare for instructors to be able to interact with students in relation to the contents of the training. As such, the most effective way of measuring the results was to make reflections on the performance of students and this has been done by many researchers [1, 24]. Table 2 below compares performance of students in the selected module in both academic years 2021 and 2022. However, it cannot be conclusively argued that the improved performance in the academic year 2022 came exclusively because of the EI interventions as other factors might have also contributed to this type of performance including student individual learning styles [1, 13, 55].

Table 2. Comparison of performance in academic years 2021 and 2022 for the chosen module

Description	2021	2022
Pass Rate	70%	100%
Distinctions	50%	79%
Pass	20%	21%
Fail	30%	0%

Nevertheless, the many positive reactions of the participants to the EI interventions as reported in Sect. 4.1 also provide evidence that this type of performance could be expected. The learning instances of the participants and behaviors exhibited by the participants can also be attributed to the type of performance as indicated in Table 2.

Considering the specific questions stated by Kirkpatrick and Kirkpatrick [27] for level 4, and reactions of the participants to such questions, overall, our participants expressed their happiness even after the training. Eleven occurrences were identified in eight participants. P1 specifically remarked: *"It was okay also to do the presentations before we do the demo, but I feel like the more effective ones were the ones we had to do the presentations in class, everyone got to learn at most"*. P2 said: *"No, I love that approach, it works 100%... no, there was no point where I felt it's too much [content] now. No, it was just enough. I don't know if it's because maybe I like programming... I don't remember anything I disliked with the explicit instruction interventions. I think I was satisfied with it"*. P3 remarked: *"I think it [explicit instruction] should be more widely implemented by other lecturers as well... I think that many lecturers give you an explanation of something without showing how it's supposed to be done. So, I think it should be also implemented in other modules. Also, studying this module was very interesting, because at Honours level, I didn't expect it to be so easy, you know what you're supposed to do"*. P5 remarked: *"I don't think there's something I really didn't enjoy, because everything was for my benefit at the end... I always tell other students that [my decision to do the] mobile programming module was the best decision I made because I was able to learn Java. Because earlier on if you asked me, if I would consider a developing job, I would say no. But the way you presented the module, and the way it was, it made me confident that I can do it. You empowered us to try and understand and figure things out by [ourselves] ... and I was able to learn a lot"*.

6 Conclusions and Future Work

CS programming instructors use various ways of teaching. As such, they continue to debate on the effectiveness of these ways. Evaluating the EI interventions in teaching computer programming can be essential in helping computer programming teaching and learning stakeholders to be aware of effective teaching strategies as well as informing how improvements can be made in the future teaching of computer programming to create a better sense of more quality and robustness. By focusing on the Kirkpatrick training evaluation model, this study aimed to: (1) provide an overview of this model, (2) relate the four levels of this model to the teaching of computer programming and (3) evaluate the use of EI interventions as a teaching approach through the model [Integrating the elements of EI with the model]. Thematic analysis of data collected through asking questions revealed that the participants in this study found a lot of value from the EI interventions, and hence appreciate such interventions. We were further able to link some aspects of computer programming teaching to the four levels of the Kirkpatrick training evaluation model. Overall, the findings of this study are consistent with the findings of many researchers [25, 44] that EI is one of the effective teaching and learning strategies, not only in other disciplines, but in computer programming as well. As we used the traditional Kirkpatrick training evaluation model, further research is needed to investigate how computer programming instructors can set themselves "*apart from and ahead of the crowd by using the four levels upside down*" as suggested by Kirkpatrick and Kirkpatrick [29, p. 8] relating to the renewed model called "The New World Kirkpartic Model".

References

1. Alturki, R.A.: Measuring and improving student performance in an introductory programming course. Informatics in Education **15**(2), 183–204 (2016). https://doi.org/10.15388/infedu.2016.10

2. Archer, A., Hughes, C.: Explicit instruction: Effective and efficient teaching. Guilford Press (2011)

3. Asghar, M.Z., Afzaal, M.N., Iqbal, J., Waqar, Y.: Evaluation of In-Service Vocational Teacher Training Program: A Blend of Face-to-Face, Online and Offline Learning Approaches. Sustainability **14**(21) (2022)

4. Baser, M.: Attitude, gender and achievement in computer programming. Middle East J. Sci. Res. **14**(2), 248–255 (2013). https://doi.org/10.5829/idosi.mejsr.2013.14.2.2007

5. Bates, R.: A critical analysis of evaluation practice: the kirkpatrick model and the principle of beneficence. Eval. Program Plann. **27**(3), 341–347 (2004). https://doi.org/10.1016/j.evalprogplan.2004.04.011

6. Bocquillon, M., Gauthier, C., Bissonnette, S., Derobertmasure, A.: Enseignement explicite et développement decompétences: antinomie ou nécessité? Formation et Profession **28**(2), 3–18 (2020) https://doi.org/10.18162/fp.2020.513

7. Bosse, Y., Gerosa, M.A.: Difficulties of programming learning from the point of view of students and instructors. IEEE Lat. Am. Trans. **15**(11), 2191–2199 (2017). https://doi.org/10.1109/TLA.2017.8070426

8. Boustedt, J., et al.: Threshold concepts in Computer Science: Do they exist and are they useful? ACM SIGCSE Bulletin **39**(1), 504–508 (2007). https://doi.org/10.1145/1227504.1227482

9. Brainscape Academy: The pros and cons of explicit grammar instruction when learning a language (2023). https://www.brainscape.com/academy/pros-cons-explicit-grammar-instruction/

10. Bretz, F.: How to Master Kirkpatrick model of training evaluation (2018). https://kodosurvey.com/blog/how-master-kirkpatrick-model-training-evaluation

11. Cahapay, M.: Kirkpatrick Model: Its limitations as used in higher education evaluation. Int. J. Assessment Tools in Edu. **8**(1), 135–144 (2021). https://doi.org/10.21449/ijate.856143

12. Cantillon, P., Hutchinson, L., Wood, D.: ABC of Leanring and Teaching in Medicine. BMJ Publishing Group, LONDON (2003)

13. Cheah, C.S.: Factors contributing to the difficulties in teaching and learning of computer programming: a literature review. Contemporary Educational Technology **12**(2), 1–14 (2020). https://doi.org/10.30935/cedtech/8247

14. Chu, X., Ilyas, I.F., Krishnan, S., Wang, J.: Data cleaning: overview and emerging challenges. In: Proceedings of the 2016 International Conference on Management of Data, pp. 2201–2206. ACM, New York (2016). https://doi.org/10.1145/2882903.2912574

15. Clark, D.: Kirkpatrick's Four Level Evaluation Model. (2015). http://www.nwlink.com/~donclark/hrd/isd/kirkpatrick.html

16. Creswell, J.W., Creswell, J.D.: Research design: Qualitative, quantitative, and mixed methods approaches, 5th ed. Sage (2017)

17. Embi, Z.C., Neo, T., Neo, M.: Using Kirkpatrick ' s Evaluation Model in a Multimedia-based Blended Learning Environment. J. Multim. Info. Sys. **4**(3), 115–122 (2017). https://doi.org/10.9717/JMIS.2017.4.3.115

18. Gardeli, A., Gardeli, A., Vosinakis, S.: Creating the computer player: an engaging and collaborative approach to introduce computational thinking by combining 'unplugged' activities with visual programming. Italian J. Edu. Technol. **25**(2), 36–50 (2017). https://doi.org/10.17471/2499-4324/910

19. Greene, K.: Understood. How to Teach Using Explicit Instruction (2023). https://www.understood.org/articles/en/how-to-teach-using-explicit-instruction

20. Grover, S., Pea, R.: Computational Thinking in K-12: A Review of the State of the Field. Educ. Res. **42**(1), 38–43 (2013). https://doi.org/10.3102/0013189X12463051

21. Guilmois, C., Popa-Roch, M., Clément, C., Bissonnette, S., Troadec, B.: Effective numeracy educational interventions for students from disadvantaged social background: a comparison of two teaching methods. Educ. Res. Eval. **25**(7–8), 336–356 (2020). https://doi.org/10.1080/13803611.2020.1830119

22. Hammond, L.: Explainer: what is explicit instruction and how does it help children learn? (2019). https://theconversation.com/explainer-what-is-explicit-instruction-and-how-does-it-help-children-learn-115144

23. Hanks, B., Fitzgerald, S., McCauley, R., Murphy, L., Zander, C.: Pair programming in education: A literature review. Comput. Sci. Educ. **21**(2), 135–173 (2011)

24. Hsiao, T.C., Chuang, Y.H., Chen, T.L., Chang, C.Y., Chen, C.C.: Students' performances in computer programming of higher education for sustainable development: the effects of a peer-evaluation system. Frontiers in Psychology, 13 (2022). https://doi.org/10.3389/fpsyg.2022.911417

25. Hughes, C.A., Morris, J.R., Therrien, W.J., Benson, S.K.: Explicit instruction: historical and contemporary contexts. Learn. Disabil. Res. Pract. **32**(3), 140–148 (2017). https://doi.org/10.1111/ldrp.12142

26. Iain, M.: ESL Lesson Handouts. Some Advantages and Disadvantages of Explicit Grammar Instruction in EFL. (2023). https://www.esllessonhandouts.com/some-advantages-and-disadvantages-of-explicit-grammar-instruction-in-efl/

27. Kirkpatrick, D.L., Kirkpatrick, J.D.: Evaluating Training Programs: The Four Levels, 3rd Edition. Berrett-Koehler (2006)

28. Kirkpatrick, J.D., Kirkpatrick, W.K.: Kirkpatrick's Four Levels of Training Evaluation (illustrate). Association for Talent Development (2016)
29. Kirkpatrick, J., Kirkpatrick, W.: An introduction to the new world Kirkpatrick model. Krikpatrick Partners (2021). http://www.kirkpatrickpartners.com/Portals/0/Resources/WhitePapers
30. Klenke, K.: Qualitative Research in the Study of Leadership. In: Klenke, K. (ed.) 2nd ed. Emerald Group Publishing Limited (2016)
31. Kunkle, W.M., Allen, R.B.: The impact of different teaching approaches and languages on student learning of introductory programming concepts. ACM Trans. Comp. Edu. **16**(1), 1–26 (2016). https://doi.org/10.1145/2785807
32. Lahtinen, E., Ala-Mutka, K., Järvinen, H.: A study of the difficulties of novice programmers. In: Proceedings of the 10th Annual SIGSCE Conference on Innovation and Technology in Computer Science Education, pp. 14–18. (2005). https://doi.org/10.1145/1151954.1067453
33. Lye, S.Y., Koh, J.H.L.: Review on teaching and learning of computational thinking through programming: What is next for K-12? Comput. Hum. Behav. **41**, 51–61 (2014). https://doi.org/10.1016/j.chb.2014.09.012
34. Marshall, C., Rossman, G.B.: Designing qualitative research, 6th ed. Sage Publications Inc. (2016)
35. Martin, R.A., Shafer, L.H.: Providing a framework for effective software quality measurement: making a science of risk assessment. In: The 6th Annual International Symposium of INCOSE: Systems Engineering: Practices and Tools, pp. 1–8 (1996). https://doi.org/10.1002/j.2334-5837.1996.tb02136.x
36. Morrison, M.J.: ABC of learning and teaching in medicine: Evaluation. BMJ **326**(7385), 385–387 (2003). https://doi.org/10.1136/bmj.326.7385.385
37. Njah, J., et al.: Measuring for success: Evaluating leadership training programs for sustainable impact. Annals of Global Health **87**(1), 1–10 (2021). https://doi.org/10.5334/aogh.3221
38. Parcell, E.S., Rafferty, K.A.: Interviews, recording and transcribing. In: Allen, M. (ed.) The SAGE Encyclopedia of Communication Research Methods. Sage Publications Inc. (2017). https://doi.org/10.4135/9781483381411.n275
39. Patton, M.Q.: Utilization-focused evaluation: the new century text, 3rd ed. Sage (1997)
40. Patton, M.Q.: Qualitative research & evaluation methods: Integrating theory and practice, 4th ed. Sage Publications (2015)
41. Peters-Burton, E.E., Stehle, S.M.: Developing student 21st Century skills in selected exemplary inclusive STEM high schools. Int. J. STEM Edu. **1**, 1–15 (2019)
42. Plowright, D.: Using mixed methods: Frameworks for an integrated methodology. Sage Publications (2011)
43. Porter, L., Garcia, S., Glick, J., Matusiewicz, A., Taylor, C.: Peer Instruction in Computer Science at Small Liberal Arts Colleges. In: ITiCSE '13: Proceedings of the 18th ACM Conference on Innovation and Technology in Computer Science Education, pp. 129–134 (2013)
44. Rastle, K., Lally, C., Davis, M.H., Taylor, J.S.H.: The Dramatic impact of explicit instruction on learning to read in a new writing system. Psychol. Sci. **32**(4), 471–484 (2021). https://doi.org/10.1177/0956797620968790
45. Reio, T.G., Rocco, T.S., Smith, D.H., Chang, E.: A Critique of kirkpatrick's evaluation model. New Horizons in Adult Education & Human Resource Development **29**(2), 35–53 (2017)
46. Rosenshine, B., Stevens, R.: Teaching functions. In: Wittrock, M.C. (ed.) Handbook of research on teaching, 3rd ed., pp. 376–391. Macmillan (1986)
47. Salleh, N., Mendes, E., Grundy, J.: Empirical studies of pair programming for CS/SE teaching in higher education: a systematic literature review. IEEE Trans. Software Eng. **37**(4), 509–525 (2011). https://doi.org/10.1109/TSE.2010.59

48. Savage, S., Piwek, P.: Full report on challenges with learning to program and problem solve: an analysis of first year undergraduate Open University distance learning students' online discussions (2019)

49. Scherer, R., Siddiq, F., Sánchez Viveros, B.: A meta-analysis of teaching and learning computer programming: Effective instructional approaches and conditions. Computers in Human Behavior **109**(0318) (2020). https://doi.org/10.1016/j.chb.2020.106349

50. Sun, D., Ouyang, F., Li, Y., Zhu, C.: Comparing learners' knowledge, behaviors, and attitudes between two instructional modes of computer programming in secondary education. Int. J. STEM Edu. **8**(54), 1–15 (2021). https://doi.org/10.1186/s40594-021-00311-1

51. Thuné, M., Eckerdal, A.: Analysis of Students' learning of computer programming in a computer laboratory context. Eur. J. Eng. Educ. **44**(5), 769–786 (2019). https://doi.org/10.1080/03043797.2018.1544609

52. TrainingCheck.com: How to Evaluate Training Effectiveness Using TrainingCheck – At a Glance Guide (2022). https://www.trainingcheck.com/help-centre-2/guide-to-training-evaluation/how-to-evaluate-training-effectiveness-using-trainingcheck-at-a-glance-guide/

53. Tshukudu, E., Jensen, S.A.M.: The role of explicit instruction on students learning their second programming language. In: UKICER '20: United Kingdom & Ireland Computing Education Research Conference, pp. 10–16 (2020). https://doi.org/10.1145/3416465.3416475

54. Wilson, B.C.: A study of factors promoting success in Computer Science including gender differences. Comput. Sci. Educ. **12**(1–2), 141–164 (2002). https://doi.org/10.1076/csed.12.1.141.8211

55. Zhang, X., Zhang, C., Stafford, T.F., Zhang, P.: Teaching introductory programming to IS students: the impact of teaching approaches on learning performance. J. Inf. Syst. Educ. **24**(2), 147–155 (2013)

Reducing Contention in an Undergraduate Capstone Project Allocation System

Stephen Phillip Levitt$^{(\boxtimes)}$ and Kenneth John Nixon

University of the Witwatersrand, Johannesburg, South Africa
{stephen.levitt,ken.nixon}@wits.ac.za

Abstract. An online project bidding and allocation system used by the School of Electrical and Information Engineering, at the University of the Witwatersrand, is described and critically analysed. An important question that arose when designing the bidding system for projects was how to deal with the contention for highly popular projects. An approach, common in the literature, is to optimise the project allocation; however, the system implemented aims to incentivise students to consider less popular projects. Projects are allocated to groups using a quasi-random algorithm, as opposed to an optimal algorithm. Over the eleven year analysis period the percentage of first choices awarded has ranged from 57.5% to 80.8%, while the average number of unallocated groups per year is 8.2%. In addition to these outcomes project contention is shown to reduce during bidding window period in nine out of the eleven years of the study. This is due to novel aspects of this system including the transparency of the allocation algorithm and the fact that groups are able to see all competing bids for the projects they are bidding on and change their own bids in response. The allocation algorithm is, by necessity, suboptimal in order to achieve the goals of transparency and fairness in that the stated winning probability, or a greater probability, for a bid always holds true when a project is allocated.

Keywords: capstone project · student project selection · allocation algorithm

1 Introduction

The Electrical/Information Engineering Laboratory is a final-year capstone course in the undergraduate degree programme of the School of Electrical and Information Engineering at the University of the Witwatersrand. The course is structured so that each academic member of staff, termed "supervisor" from here onwards, offers a small number of unique projects (between one and three). Students are required to work in pairs (as teamwork is one of the outcomes being assessed) and must undertake one particular project on offer.

Prior to the introduction of the system the allocation of projects to student groups was done in a manual fashion. This manual allocation process was

H. E. Van Rensburg et al. (Eds.): SACLA 2023, CCIS 1862, pp. 34–47, 2024.
https://doi.org/10.1007/978-3-031-48536-7_3

generally perceived as being unfair, especially in cases where there were highly contested projects, because there were no clear guidelines to students, and supervisors, as to how projects were being allocated. An additional issue was that this manual process was becoming unwieldy with increasing student numbers.

Given the importance of this course and the level of unhappiness that students were feeling at the time, the authors undertook to completely redesign and automate the project allocation process with the goals of still allowing student agency (in choosing projects) while promoting system transparency (in that all actions taken with regard to the allocation process are visible to all parties involved). A key consideration in the design was how to deal with the contention for highly popular projects. Rather than adopting the approach of optimising the project allocation, the solution adopted aims to incentivise students to consider less popular projects. This new system, and the way in which the above goals are achieved, is described in the following sections. Section 2 presents the relevant literature. Section 3 concentrates on the online bidding process, while Sect. 4 discusses how projects are allocated. In Sect. 5 analyses of various aspects of the system are provided. This is followed by the conclusion.

2   Existing Approaches

There are three key entities in student-project allocation problems which are *students*, *projects*, and *supervisors*. Approaches to allocating projects vary by allowing different kinds of constraints among the key entities. Some approaches solely allow students to rank projects in order of preference [10, 15]. Others may additionally allow supervisors to specify their preferences over students, as in [1]. Supervisors may also specify their preferences over projects [12]. Finally, a fairly generalised approach is given wherein supervisors specify preferences with regard to which students they want, on the projects they want, and students specify their preferences with regard to the projects [7, 13]. Many approaches also satisfy capacity constraints with regard to the number of projects that a supervisor may take on, and the number of students that may be allocated to a particular project.

Student-project allocation problems can be seen as specific instances of generalised assignment problems for which a body of mathematical and computational theory is applicable [4]. These problems are shown to be NP-hard in complexity, and more recently, certain formulations have been proven to be APX-hard [11, 12] which means that approximate solutions within known bounds are attainable within polynomial-time.

A number of different techniques for solving these problems have been described. These include ad-hoc algorithms [15], integer programming [2, 5], genetic algorithms [10, 14], and two-sided matching algorithms including the classic Gale-Shapley stable matching algorithm [8] used by [9] and variants of this [1, 7, 12].

In arriving at a solution an objective function is defined and a search is conducted for an allocation which maximises this function. An optimal solution is often seen as one in which students are allocated their highest ranked projects as far as possible whilst minimising the number of unallocated students. In the case of two-sided matching problems, wherein, for example, both students and supervisors express preferences relating to one another, a stable matching is sought in which the number of unallocated parties is minimised. A stable matching is one in which no two parties who are not matched together would rather be matched to each other than their current assignees.

Existing literature on the student-project allocation problem has tended to focus narrowly on contributing to the theoretical understanding of the computational complexity of this type of assignment problem and it variants, and presenting algorithms for finding approximate or optimal solutions. The starting position, of much existing work, is seen to be a predetermined set of preferences and constraints, and the goal is to optimise the allocation given these initial conditions.

However, there are a few studies that specifically consider the broader social context in which the student-allocation problem is situated. In the work done by Greef et al. [9] an interesting approach is taken whereby students tender for projects, and express their project preferences by ranking their own tenders. They further participate in ranking the tenders of their peers. These rankings are combined and form the preference scores that are used by the stable matching algorithm. This promotes student engagement with the system and students are exposed to anonymised, competing tenders.

Briffa et al. [3] discuss an allocation system which makes use of a web application to provide real-time feedback to students concerning the popularity of topic choices. In this way, their system allows students to opt for less popular topics if they are at risk of not receiving their top preferences. They use a simulated annealing algorithm to seek a global optimum topic allocation.

The approach which is presented below is closest to those of [10, 15], in that students express preferences over projects and there are no other constraints involved. However, a deliberate decision is taken to include the broader social context within which the project allocation is occurring in a somewhat similar manner to [3], and not to solely focus on an algorithmic solution to a set of expressed preferences.

3   Project Bidding

In the Electrical/Information Engineering Laboratory, students initially pair up by mutually selecting each other using the online system. Students are then given one week to obtain bidding rights followed by another week in which the actual bidding for projects takes place.

3.1   Obtaining Bidding Rights

Obtaining bidding rights is a important step which requires each group to meet with potential supervisors in order to discuss the projects that they are interested in. This ensures that groups have, at least, an initial engagement with supervisors regarding the requirements and expectations of projects that they are interested in. The supervisor, in turn, is obligated to grant the group bidding rights on these projects after this discussion. This will allow the group to place bids on these projects once the bidding window opens. Groups may also continue to obtain bidding rights on projects after the bidding window has opened.

Supervisors may not selectively grant bidding rights to groups that they would prefer. Unlike some other systems that have been documented in the literature, this system does not allow for supervisors to express preferences over students. This decision has been taken by the School because it supports the notions of fairness, in that supervisors may not prefer (cherry-pick) certain groups over others, and quality, in that any group in the final-year cohort, should be capable of doing on any project on offer that they may be interested in. One of the drawbacks of this decision is that it could potentially lead to supervisors being marginalised if students consistently avoid bidding on any of their projects. In practice, this has been found to be a minor issue with students only tending to avoid supervisors that they have never met before in a lecturing capacity. Within our School, this has applied to very few supervisors over the years and, in such cases, deliberate interventions can be put in place to introduce "unknown" supervisors to students.

All students and supervisors are able to see the number of groups that have obtained bidding rights on any particular project. This gives an idea as to the initial popularity of the project and is intended to encourage students to obtain bidding rights on additional projects if they see that all of the projects that they are interested in are popular.

3.2 Placing Bids

Each project has six bidding slots available. There are three first-choice slots, two second-choice slots, and one third-choice slot. This allows a maximum of six bids from six different groups for each project. Of course, projects may receive fewer bids or no bids at all.

Once the bidding window opens, each group must submit bids for three different projects *simultaneously*: a first-choice bid, a second-choice bid, and a third-choice bid. A group can only bid on projects for which they have obtained bidding rights. Furthermore, each project must have an open bidding slot for the particular choice of bid that they wish to make. The single third-choice slot is the limiting factor here, and groups often find that they need to secure bidding rights on additional projects because the third-choice slots for the projects on which they do have rights are already occupied.

The system guarantees that the probability of a first-choice bid winning a particular project is always greater than the probability of a second-choice bid, and that the probability of a second-choice bid winning a particular project is always greater than the probability of a third-choice bid. This is discussed in more detail in Sect. 4.

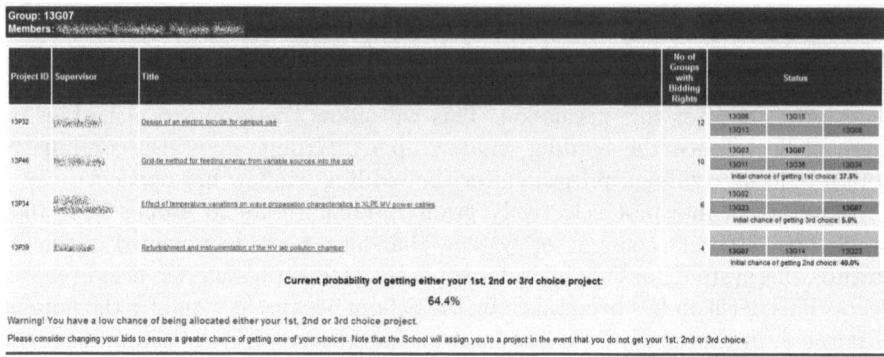

Fig. 1. Student view of bidding status (anonymised). This particular group has bidding rights on four projects and have placed bids for the bottom three projects in the list.

Groups may repeatedly change their bids throughout the bidding period. The system is entirely transparent in that students can see exactly which other groups are bidding for the projects that they are interested in. They can also see their probability of being awarded a project that they are making a bid on, assuming that it is processed independently of all the other projects. Finally, their overall probability of being allocated any of the projects that they have bid on by the allocation algorithm is calculated using the per-project probabilities. If the overall probability is low for the group then they are issued with a warning that they may not, in fact, be allocated any project by the system. This is illustrated in Fig. 1.

After a week of bidding, the bidding window is closed and the projects are then allocated using an automated allocation algorithm (coded in Octave [6]) which is described in Sect. 4. The source code of the algorithm is made available for students to download, along with instructions on how to run the algorithm against the current bidding state. In addition to this, *throughout the bidding period* students are able to run the algorithm directly from the course website and view the output (which varies due to its inherent randomness). The algorithm output is detailed, showing not only the final outcome of the allocation, but also clearly documenting each step that is taken in reaching those results.

If any group fails to make their bids within the bidding period they are allocated a project manually, from the remaining, unallocated projects.

4   Project Allocation

In order to allocate a given project, one of the bids for the project is randomly selected by the allocation algorithm and the group that placed that bid wins the project. Not all of the bids stand an equal chance of being selected and this is described in more detail below.

Table 1. Processing Order for Project Categories

Sequence	Project category (based on bids present)	Bidding pool selection probability
1	First-choice bids only	100 : - : -
2	First- and second-choice bids only	79 : 21 : -
3	First-, second- and third-choice bids	75 : 20 : 5
4	Second-choice bids only	- : 100 : -
5	Second- and third-choice bids only	- : 80 : 20
6	Third-choice bids only	- : - : 100

4.1 Weighted Bidding Pool Probabilities

All the first-choice bids for a particular project form the first-choice pool. All the second-choice bids form the second-choice pool and the single third-choice bid forms the third-choice pool. If all three bidding pools have bids present, then the probabilities are weighted such that the first-choice pool has a 75% probability of being selected, the second-choice pool, a 20% probability, and the third-choice pool a 5% probability. In other words, the likelihood that a bid in a particular pool will be selected can be expressed using the following ratios:

$$\text{First-Choices : Second-Choices : Third-Choice}$$
$$15 \quad : \quad 4 \quad : \quad 1$$

A first-choice bid is fifteen times more likely to be selected by the algorithm than a competing third-choice bid.

In cases where a bidding pool is not occupied, the empty pool is omitted from selection. The ratios given above are preserved for the remaining bidding pools by redistributing the empty pool's probability of selection proportionately. In cases where there are two unoccupied pools, the remaining pool will have a 100% probability of being selected. The probability of each bidding pool being selected in these different scenarios is illustrated in Table 1.

4.2 Project Processing Order

A critical aspect of the allocation algorithm is the order in which the projects are allocated to groups. In order to honour the ranking of choices it is essential to process projects, for example, which have first-choice bids before projects that only have third-choice bids. The projects are therefore grouped into categories based on the type of bids present. The category are processed in the order shown in Table 1.

Projects are allocated in sequence starting with those belonging to the first category listed in Table 1. If there is more than one project in any category, then one of the projects in the category is selected randomly and the project allocated. Once a project is allocated it is removed from the project list. All of the remaining bids of the "winning group" are deleted from the project list, the probabilities for each project are recalculated, and the list is re-categorised. Project allocation continues in this fashion until there are no more projects remaining in the list.

If a group makes all of its bids for contested projects there is a chance that they may not be allocated any of their choices. In other words, competing bids may be selected for each of the projects on which they have bid. If this occurs the group concerned is manually allocated a project (in consultation with the course co-ordinator) from those that remain after the allocation has been finalised. In spite of being warned about this consequence, some groups decide to take the risk for the chance of doing a project that they are really interested in and they enter the final allocation with a low overall chance of being allocated a project by the system.

There may be multiple first and second-choice bids for a project. The probability of any *one of the bids* winning the pool is equal to the bidding pool probability divided by the number of bids in the pool. So, for example, assume that a project has two first-choice bids, two second-choice bids and one third-choice bid present. The probability of either of the first choices winning is 37.5%, the probability of either of the second choices winning is 10%, and the probability of the third choice winning is 5%.

4.3   Allocating a Project

Selecting a winning bid from the bidding pools for any given project is readily implemented. An array representing the bids for the project is created. Seventy-five elements of the array are used to represent the first choice bidding pool, twenty the second-choice bidding pool, and five the third-choice.

In the scenario where there is one first-choice, one second-choice and one-third choice bid for the project, all seventy-five elements representing the first choice will contain the identifier of the group which made the project their first-choice, twenty will contain the identifier for the group making the project their second-choice, and five will contain the identifier of group making it their third-choice. In order to allocate the project, one of the array elements is randomly selected.

In cases where there are multiple groups within a single bidding pool (first- and/or second-choice bids) the array elements are divided evenly among the bidding groups. For example, if there happen to be three first-choice bids then each of the bidding groups will be allocated twenty-five elements of the array. Likewise, if there are two second-choice bids then each the groups will be allocated ten elements.

The array is sized to only represent bidding pools which have bids present. For example, if there are no third-choice bids, the array will consist of only ninety-five elements with seventy-five being reserved for first-choice bids and twenty being reserved for second-choice bids. As before, after the array is populated with the bidding groups an element is randomly selected to determine which group is given the project.

4.4   Finalising the Allocation

Owing to the deliberate random nature of the allocation algorithm, it may happen that a single run of the algorithm produces an unfavourable result. An unfavourable result would be where a relatively large number of groups are not allocated a project or the number of first choice allocations is relatively low. To deal with this possibility, the allocation algorithm is run three times in the presence of the student class representatives, the course co-ordinator, and a senior administrative officer. The class representatives then select one of the three runs as the final allocation. In doing this they need to decide which of the runs represents the best trade-off between honouring student preferences and minimising the number of unallocated groups. At no time are they privy to which groups have received which projects as this could bias their decision. Once the final allocation has been decided upon it is published to the class, along with the entire output of the selected run.

5   System Performance and Analysis

In analysing the system that has been presented, the statistics are given first which provide insight into various aspects of the allocation algorithm's performance over an eleven year period. This is followed by a detailed discussion of the two different ways in which this system reduces the contention for projects. Lastly, the optimality of the allocation algorithm is considered.

5.1   System Statistics

Table 2 presents the statistics for the system that have been gathered since its inception. The system's components were not changed over this time period and therefore these yearly statistics are comparable with each other. The first column designates the year. The second column indicates the number of groups that were involved in the bidding process. This is followed by the number of projects that were available for students to bid on, and the number of excess projects (total projects on offer minus groups bidding) expressed as a percentage. The next two columns show the outcome of the allocation algorithm, in particular the number of winning first-choice bids (column 5) and the number of unallocated groups (column 6) expressed as percentages.

It is clear from the table that this system is able to grant the majority of groups their first choices and this is a direct result of using weighted bidding pool probabilities. Looking at the results in more detail, it is evident that Year 3 is an outlier in terms of the number of groups bidding. The forty groups bidding in that year is far higher than any of the other years. Additionally, the number of excess projects in that year (7%) was the smallest. So, essentially, year 3 represented a stress test for the system in that one would expect a high amount of contention for projects. The allocation results show that the system was still able to grant a majority of first choices at the expense of a relatively high number

Table 2. Statistics for the System

Year	Groups bidding	Available projects	Excess projects (%)	First choices granted (%)	Groups unallo- cated (%)
1	23	39	41.0	60.9	4.4
2	28	37	24.3	62.1	10.7
3	40	43	7.0	57.5	20.0
4	26	29	10.3	80.8	11.5
5	25	29	13.8	72.0	8.0
6	31	46	32.6	71.0	9.7
7	25	40	37.5	73.1	4.0
8	27	40	32.5	63.0	0.0
9	24	39	38.5	75.0	0.0
10	27	31	12.9	63.0	7.4
11	27	34	20.6	59.3	14.8

of unallocated groups (20%). In other years, with fewer groups, the percentage of unallocated groups is far less.

This suggests that the system is unable to deal effectively with larger group numbers if there are insufficient excess projects. From the table, it appears that the number of excess projects needs to be at least greater than 10% in order to keep the number of unallocated groups small. Where there is a larger number of extra projects (years 6 through 9) it is possible to grant a high number of first choices with few unallocated groups, in some cases zero unallocated groups. Of course, the effort in this regard needs to be balanced because it entails more work on the part of the supervisors and coordinators who have to propose and review additional projects.

The allocation system appears to have been well received by students from the informal feedback that has been elicited over the decade that the system has been employed. There have been less than a handful of complaints and these have originated from groups that have tried to game the system. Instances of attempting to game the system have occurred during the bidding process when a group has observed the outcome of a number of algorithm runs and seen that in these runs they do not receive their first-choice project, so in order to try and rectify this they change their first-choice bid to a second-choice bid and (mistakenly) hope this will improve their chances.

It is arguable, that the success of the system lies not in its ability to award the majority of groups their first-choice projects, which it does, but rather in that it is transparent and accustoms students to the idea that they may not receive their first-choice project.

5.2   Project Contention

The level of projection contention is a crucial factor for any technique which attempts to solve the student-project allocation problem. The more highly contested projects are, the more disappointing the results of the allocation will be for the majority of students irrespective of which allocation technique is adopted and whether or not the allocation is optimal, according to the objective function, or sub-optimal (approximate). This simply attests to the fact that many groups/students in highly-contested scenarios will not be able to receive any of their top-ranked projects.

Another issue that is of concern, is that groups may in fact not receive any projects at all after the allocation process has run, having lost to competing groups on all of the projects that they had ranked. Different approaches to handling this situation are documented in the literature. In [15], for example, unassigned groups re-rank the remaining, unassigned projects. In [10] students may be a assigned to a "non-choice" which is a project that they did not actually rank. A solution to this is to increase the number of projects that students must initially rank in order to give the allocation algorithm a greater possibility of finding an allocation in which all students receive one of their preferences.

This system deals with project contention in two different ways. The first is structurally, and the second is through the social dynamics at play.

Reducing Contention Through Bidding Slot Structure. The bidding slots have been structured in the manner described above (three first-choice slots, two second-choice slots and one-third choice slot per project) limiting the contention for any one project to a maximum of six different groups.

The motivation for such a structure is twofold. On the one hand, popular projects exist and it is desirable to have a system which takes cognisance of this by allowing multiple bids, in particular first-choice bids. If the system was not structured in this manner, and there was, for example, only a single first choice slot for projects, then whichever group was the quickest in making their bids at the opening of the bidding session would most likely receive the project. It is not the intention that projects are awarded on a "first-come, first-served basis". Given the total number of groups in the School, having three first choices gives groups a reasonable chance of making a first-choice bid on the projects that they are interested in.

On the other hand, in order to have viable system in which most groups are allocated projects, it is necessary to force groups to spread their bids. The "bidding spread" can be defined as the number of projects which have at least one bid. The spread when considering only first choice bids, in the worst-case, maximally-contested, scenario (smallest spread) will be equal to the number of groups divided by three. However, for the third-choice bids the spread will always equal number the groups, which is the widest spread possible. The two second-choice slots enforce an intermediate spread. Having such a structure leads to the following bounds on the average number of bids per project:

$$\text{Average bids per project} = \frac{\text{total bids}}{\text{number of projects bid upon}}$$

$$= \frac{\text{groups bidding} \times 3}{\text{number of projects bid upon}}$$

Table 3. Bidding System Social Dynamics

Year	Bidding state changes (all bidding)	Change in project contention (all bidding)
1	11	0.000
2	28	−0.071
3	142	−0.045
4	192	−0.037
5	49	−0.060
6	56	0.000
7	15	−0.056
8	14	−0.040
9	46	−0.020
10	19	−0.058
11	78	−0.073

The upper bound of the average is thus equal to three because the number of projects that are bid upon must, at a minimum, be equal to the number of groups bidding as each group is required to make a third-choice bid and there is only one such slot per project. So although the maximum number of bids for any one project is six bids, the average bids per project will, at most, be three which is low.

Reducing Contention Through Social Interaction. Given the transparent nature of the system, in that groups can see competing bids, it is interesting to examine the social interaction which is at play. Two metrics that provide some insight into the bidding behaviour are state and contention changes as detailed in Table 3. Groups often desire and compete for the same projects; however, they run the real risk of not being awarded any projects if they only place bids on popular projects. It is therefore reasonable to assume that, on the whole, the class will attempt to self-optimise by moving away from popular projects to some extent.

In order to examine this hypothesis, it is necessary to analyse the bidding activity that takes place. The time period of interest for this analysis begins at the point at which all groups are on board and have made bids and ends at the close of bidding. This particular time span is chosen for two reasons. Firstly, as groups join the bidding over time they introduce "turbulence" into the system by supplying three completely new bids and altering the bidding landscape. Therefore, by only considering the period from which all groups have bids present this effect is negated and the students are able to optimise against a relatively stable system state. The second reason is that students at the tail-end of the bidding period are forced to "show their hand" before the bidding closes so there should be less attempts to play the system and a concerted effort to try and improve their chances of receiving a project.

The number of bidding state changes which occurred during this time span, over the years since the system's inception can be seen in Table 3. The bidding state changes whenever a group updates its bids. The simultaneous submission of all three of the group's bids (a first, second and third choice) is counted as a single state change. It is evident that there is a fair amount of activity that takes place as groups manoeuvre in response to each other's bids but there does not appear to be a correlation with the total number of groups that are bidding.

Lastly, in order to support this analysis a *project contention measure* is defined as being the total number of bids that have been placed divided by the total number of occupied bidding pools (bidding pools which have at least one slot occupied). This measure attempts to capture how widely students are spreading their bids. If the bids cluster around popular projects then fewer bidding pools will be occupied as popular projects will have many bids per pool. If the bids are widely spread among different projects than more pools will be occupied with the pools containing less bids on average.

This measure is calculated at the start and end of the time period of interest and difference is given in the last column of Table 2. In nine out of the eleven years that the system has been operation, the project contention has dropped by the close of bidding. For remaining two years (years 1 and 6), it stayed the same. These statistics offer strong evidence that the class is self-optimising and that the social dimension to the system has a small but tangible effect in causing the bids to spread.

It is reasonable to expect that as the project contention is decreased there will be an increase in the number of first choices granted, in other words, a negative correlation. From the limited number of data points there is a weak positive correlation ($r = 0.23$) which seems counter-intuitive. However, the number of excess projects should also have an effect on the number of first choices granted. Ultimately many more data points are needed to have confidence in the correlation between the variables.

5.3   Optimal Versus Non-optimal Allocation

Given the relatively small search space it is possible to use a brute-force search, or one of the other optimisation techniques mentioned in Sect. 2, to find an optimal, or near-optimal, allocation of projects, for a given bidding state, by maximising the percentage of first-choice allocations (or any other cost function related to the system outputs).

However, this presents a conundrum in that certain groups will consistently, throughout the bidding period, be granted second or third choice projects in order for an optimal allocation to be achieved. In a sense, these groups have to be "sacrificed" for the greater good of the entire cohort. Given the transparent nature of this system with the ability to run the allocation algorithm at any point in time on the current bidding state, the inability for some groups to ever achieve their first choices would become clear and this would effect the perceived fairness of the system. Therefore, a sub-optimal algorithm is used and because of its random nature the allocation of projects is not predetermined based on the bidding state.

6   Conclusion

The student-project allocation approach that is presented here allows students to specify preferences for projects by making first-, second- and third-choice bids. No constraints of any other kind are incorporated. The deliberate random nature of the allocation algorithm produces approximate solutions for a given bidding state rather than a single optimal or near-optimal solution.

This system has a strong social dimension and is highly transparent. During the bidding process, groups are able to see competing bids on the projects that they are interested in and change their own bids in response. They are presented with their probability of being awarded a project by the system. Additionally, they are able to run the allocation algorithm from the course website and view the results, which will vary on each run due to algorithm's randomness. This results in a system that is fair, or egalitarian, in that no groups are precluded from being awarded their first choice project, which would potentially be the case for some groups if an optimal solution was generated.

Particular care has been taken to design the allocation system to reduce project contention but still accommodate popular projects. The system is structured in such a way that there are six bidding slots per project. However, the single third-choice bidding slot per project enforces groups to spread their bids widely. The social dynamics of the system are shown to further reduce project contention as groups are generally risk averse and would rather move some of their bids to less popular projects than risk not being allocated a project at all. A reduction in project contention, in turn, makes it is easier for the allocation algorithm to perform effectively. Ultimately, such a system affords students a strong sense of agency in that not only are students able to decide on which projects to bid for, they are also able to decide for themselves on the level of risk that they wish to undertake with respect to being awarded a project by the system.

References

1. Abraham, D.J., Irving, R.W., Manlove, D.F.: Two algorithms for the student-project allocation problem. J. Discrete Algorithms **5**(1), 73–90 (2007)
2. Anwar, A.A., Bahaj, A.S.: Student project allocation using integer programming. IEEE Trans. Educ. **46**(3), 359–367 (2003)
3. Briffa, J.A., Lygo-Baker, S.: Enhancing student project selection and allocation in higher education programmes. In: 2018 28th EAEEIE Annual Conference (EAEEIE), pp. 1–6. IEEE (2018)
4. Burkard, R., Dell'Amico, M., Martello, S.: Assignment Problems, Revised Reprint. Society for Industrial and Applied Mathematics (2009)

5. Chiarandini, M., Fagerberg, R., Gualandi, S.: Handling preferences in student-project allocation. Ann. Oper. Res. **275**(1), 39–78 (2019)
6. Eaton, J.W.: GNU Octave. https://octave.org/. Accessed 11 Apr 2023
7. El-Atta, A.H.A., Moussa, M.I.: Student project allocation with preference lists over (student, project) pairs. In: 2009 Second International Conference on Computer and Electrical Engineering, vol. 1, pp. 375–379. IEEE (2009)
8. Gale, D., Shapley, L.S.: College admissions and the stability of marriage. Am. Math. Mon. **69**(1), 9–15 (1962)
9. Greeff, J.J., Heymann, R., Nel, A., Carroll, J.: Aligning student and educator capstone project preferences algorithmically. In: 2018 IEEE Global Engineering Education Conference (EDUCON), pp. 521–529. IEEE (2018)
10. Harper, P.R., de Senna, V., Vieira, I.T., Shahani, A.K.: A genetic algorithm for the project assignment problem. Comput. Oper. Res. **32**, 1255–1265 (2005)
11. Iwama, K., Miyazaki, S., Yanagisawa, H.: Improved approximation bounds for the student-project allocation problem with preferences over projects. J. Discrete Algorithms **13**, 59–66 (2012)
12. Manlove, D.F., O'Malley, G.: Student-project allocation with preferences over projects. J. Discrete Algorithms **6**(4), 553–560 (2008)
13. Moussa, M.I., El-Atta, A.H.A.: A visual implementation of student project allocation. Int. J. Comput. Theory Eng. **3**(2), 178–184 (2011)
14. Salami, H.O., Mamman, E.Y.: A genetic algorithm for allocating project supervisors to students. Int. J. Intell. Syst. Appl. **8**(10), 51 (2016)
15. Teo, C.Y., Ho, D.J.: A systematic approach to the implementation of final year projects in an electrical engineering undergraduate course. IEEE Trans. Educ. **41**(1), 25–30 (1998)

Factors Determining the Success of Online Learning Videos for Programming

Janet Liebenberg$^{(\boxtimes)}$ ⓘ and Suné van der Linde ⓘ

School of Computer Science and Information Systems, North-West University, Potchefstroom, South Africa
janet.liebenberg@nwu.ac.za

Abstract. With the COVID-19 pandemic causing universities to close, online learning became a popular solution for educators and students. This study explored the factors that determine the success of video-based online learning for a programming course. The programming course was designed based on principles from a Problem-Solving Learning Environment (PSLE) that develops computational thinking, with video lectures forming part of the scaffolding and information processing components. To conduct the research, a mixed methods approach was used, using a survey for quantitative data collection and open-ended questions to collect qualitative data from 509 survey respondents taking a C# programming course. The researchers used the Unified Theory of Acceptance and Use of Technology (UTAUT) as a lens in order to make sense of and obtain a deeper understanding of the factors, including performance expectancy, effort expectancy, attitude towards using technology, facilitating conditions, and behavioural intention. The use of online concept videos was found to be beneficial for learning programming concepts, with participants reporting improvements in academic achievement, increased efficiency, low effort, enjoyment, and compatibility with learning styles. Key factors for success include ease of use, short duration, relevance, thoroughness, engagement, and availability of necessary resources. The study provides insights for lecturers of programming courses to create effective online learning videos.

Keywords: Online Learning Videos · Online Teaching · Online Learning · Programming Teaching

1 Introduction

The COVID-19 pandemic has drastically changed the way education is delivered across the world [17]. With the closure of schools and universities, online learning has become an apparent solution for countless educators and students. Video-based learning has emerged as one of the most popular forms of online learning due to its ability to deliver rich multimedia content and facilitate self-paced learning [15]. Software development or programming is a highly in-demand skill and software development roles (including software engineers, full-stack developers, and front-end developers) are currently high in demand [18]. Online learning videos have gained significant popularity in recent

© The Author(s), under exclusive license to Springer Nature Switzerland AG 2024
H. E. Van Rensburg et al. (Eds.): SACLA 2023, CCIS 1862, pp. 48–63, 2024.
https://doi.org/10.1007/978-3-031-48536-7_4

years due to their ability to provide an interactive and engaging learning experience [6]. However, the success of online learning videos and moreover videos for learning programming depends on several factors [5]. In this study, we explore the factors that determine the success of online learning videos for programming and provide insights into how educators can create effective online learning videos to improve students' learning outcomes.

2 Background

2.1 Online Teaching and Learning

The Horizon Report for Higher Education presents trends, challenges, and developments in educational technologies projected annually to have an impact on colleges and universities across the United states. For universities, the integration of online, mobile, and blended learning approaches is critical to their survival, according to the 2016 and 2017 reports and the reports call for an exploration of how these models are actively enriching learning outcomes. [3, 9]. Little did the authors of those reports realize that COVID-19 will make their predictions regarding online learning approaches which are critical for survival so much more applicable. Not surprisingly after the outbreak of Covid-19, the 2022 Horizon report [17] identified the following key technologies and practices that will have a significant impact on the future of higher education institutions' teaching and learning: AI for learning analytics and learning tools, hybrid learning spaces, mainstreaming hybrid/remote learning modes, micro-credentialing and professional development for hybrid/remote teaching.

In their study, Parker and van Belle [16] found that most students indicated that technology has enabled them to develop additional skills and also be a significant factor in course-specific learning.

Fidalgo, Thormann, Kulyk and Lencastre [7] conducted a study in Portugal, UAE and Ukraine and found that in all three countries, students' major concerns about online learning were time management, motivation, and English language skills. [7] recommend that lecturers need to be trained to develop and deliver online courses that help to overcome obstacles such as time management and motivation.

In her research, Ayebi-Arthur [1] conducted a case study at a New Zealand college that faced significant disruptions due to seismic activities. Through her investigation, she discovered that the college exhibited increased resilience in adopting online learning following the disastrous event. Technology played a crucial role in surmounting the challenges they faced during those trying times. However, the study emphasizes the importance of having a sturdy IT infrastructure as a prerequisite for successful online learning. The infrastructure should be robust enough to ensure uninterrupted services both during and after crisis.

It is becoming increasingly clear that effective teaching in online courses differs from that in traditional classroom settings. Studies have revealed that teaching online fundamentally differs from teaching in person, necessitating the development of new lesson-planning techniques by lecturers [14]. There is a need for a critical and nuanced understanding of the outcomes of new technologies on the habits and subject positions of learners and teachers in higher education. In higher education, there is an urgent need for a sophisticated understanding of how new technology affect the practices and subject areas of students and lecturers [2, 11].

2.2 Online Learning Videos

The studies conducted by Mayer and colleagues [14] have identified multimedia learning principles that serve as guidelines for instructors to create effective lesson plans in a multimedia setting. These principles include managing essential overload, reducing extraneous processing, and employing social cues to improve learning outcomes. The risk of exceeding students' cognitive capacity can be minimized by creating videos with learner-paced segments, using familiar names and terms, and speaking instead of using on-screen text. To reduce distractions, course designers can eliminate extraneous information, combine narration with animation simultaneously, use cues to highlight essential information, and both words and pictures should be arranged so that they are proximal in space and time. Students learn best when spoken to in a conversational manner with a human voice.

Choe, Scuric, Eshkol, Cruser, Arndt, Cox, Toma, Shapiro, Levis-Fitzgerald and Barnes [4] identified common online video styles referred to as Classic Classroom, Weatherman, Demonstration, Learning Glass, Pen Tablet, Interview, Talking Head, and Slides On/Off. Despite having similar learning outcomes, they discovered that students show substantial preferences for particular video. Among the styles, Learning Glass received the highest satisfaction ratings. A low rating was given to video styles described as impersonal and unfamiliar, while a high rating was given to those described as engaging and personal.

Diwanji, Simon, Märki, Korkut and Dornberger [5] offered a list of online learning video success factors divided into categories. The video's style and content are important including short duration, Khan-style, personal feeling, enthusiasm, humour, high-definition quality, and graphical elements. Assessments and supplemental materials are crucial auxiliary materials. Investment in pre-production and expert assistance while recording are equally important. For a broader audience, distribution networks, social media, and internationalization via dubbing and subtitles are required. It's also key to incorporate gamification through the granting of badges and user data. Finally, it's imperative to use mobile devices and applications for videos and courses.

In the study of Manasrah, Masoud and Jaradat [13] they compared the performance of two groups of engineering students who were given either short or long lecture videos. The results showed that students who received short videos performed statistically significantly better in the course and found the videos more entertaining. However, students who did not find the videos informative scored significantly lower marks. Their study suggests that the optimum length for a lecture or tutorial video for engineering students is between 6 to 10 min, and videos shorter than 3 min were incomplete and insufficient.

In terms of the design of online learning video environments for programming teaching Elçiçek and Karal [6] identified four design variables, namely "content", "interaction", "practicability" and "visual design".

This study, therefore, aimed to determine the success factors of online learning videos within a programming course.

3 Empirical Investigation

In this section the demographics of the participants will first be explained, followed by the data collection and analysis and finally, the results will be discussed.

3.1 Settings and Participants

During this study, the North-West University (NWU) in South Africa was used as a research site. The programming course included in this study is called User Interface Programming 1 and is a part of the BSc in Information Technology degree. It focuses on using the C# programming language. The course is typically taught by two lecturers, each located at a different campus of the NWU in Potchefstroom and Vanderbijlpark. Furthermore, the course is available as a distance-learning option. The module content, assessments, and contact hours are aligned across all delivery methods.

Due to the impact of Covid-19, the course was converted into an emergency remote online course during the second semester of 2020. In this modified format, the course was taught to a total of 632 students by two lecturers. One of the lecturers handled the Potchefstroom campus students as well as the distance learners, while the other lecturer took charge of the remaining students at the Vanderbijlpark campus.

3.2 User Interface Programming 1

The learning outcomes of the course are provided in Table 1. The course contains seven study units. In the table, a code is assigned to each learning outcome with a reference to the study unit (e.g. Study Unit 1 – S1), followed by the learning outcome number (e.g. Learning Outcome 1 – L1), for example, S1L1.

The course instructional design is based on principles from Lye and Koh [12] for a Problem-Solving Learning Environment (PSLE) which develops computational thinking. The PSLE design of the online learning management system course environment incorporates authentic problems, information precessing activities based on cognitive constructivism, scaffolding and reflection. PSLE is a well-rounded approach developed by numerous prominent authors in the field, incorporating a range of perspectives. (Table 2).

The concept videos form part of the information processing and scaffolding components (Table 2). Our goals with the concept videos were to keep the videos:

- short,
- aligned with learning outcomes, and
- presented in a way to promote the mental model construction of concepts.

We aligned the videos with the course learning outcomes and created the following videos to address the key outcomes of the course. Table 3 presents the videos created, their duration and their alignment with the course outcomes.

Table 1. Course Learning Outcomes.

Study Unit (SU)	Code	Learning Outcome Description
SU1	S1L1	Demonstrate knowledge about graphical user interfaces;
	S1L2	Demonstrate knowledge about objects and classes;
	S1L3	Understand the program development process;
	S1L4	Work in the Visual Studio environment;
	S1L5	Create a GUI for a Visual C# application;
	S1L6	Write code for the Visual C# application;
	S1L7	Demonstrate the use of forms and controls including label and picture box controls;
	S1L8	Understand the use of IntelliSense, comments, blank lines, and indentation;
	S1L9	Write code to close an application's form;
	S1L10	Know how to deal with syntax errors;
SU2	S2L1	Read input with textbox controls;
	S2L2	Understand variables and numeric data types;
	S2L3	Perform calculations;
	S2L4	Input, output and format numeric values;
	S2L5	Perform simple exception handling;
	S2L6	Use named constants;
	S2L7	Declare variables as fields;
	S2L8	Use the Math class;
	S2L9	Fine-tune the GUI; and
	S2L10	Use the debugger to locate logic errors
SU3	S3L1	Write decision structures including the if, if-else and nested decision statements;
	S3L2	Understand equality, relational, and logical operators used with conditional expressions;
	S3L3	Know conditional expressions that use bool variables and flags;
	S3L4	Compare strings;
	S3L5	Execute input validation and prevent data conversion exceptions with the TryParse methods;
	S3L6	Write switch statements; and
	S3L7	Create applications that use radio buttons, check boxes and list boxes
	S3L8	Know more about ListBoxes

(continued)

Table 1. (*continued*)

Study Unit (SU)	Code	Learning Outcome Description
SU4	S4L1	Know more about ListBoxes;
	S4L2	Write while loops;
	S4L3	Use the $++$ and $--$ operators;
	S4L4	Write for loops;
	S4L5	Write do-while loops;
	S4L7	Use the OpenFileDialog and SaveFileDialog controls;
	S4L8	Use random numbers; and
	S4L9	Write a load event handler
SU5	S5L1	Write your own void methods;
	S5L2	Distinguish between passing arguments to methods by value or by reference;
	S5L3	Write your own value and nonvalue-returning methods; and
	S5L4	Debug methods
SU6	S6L1	Understand the concepts of classes and objects;
	S6L2	Create multiple forms in a project; and
	S6L3	Access a control on a different form
SU7	S7L1	Know about usability goals and measures;
	S7L2	List and shortly describe usability motivations; design;
	S7L3	Understand universal usability;
	S7L4	List and describe the goals of user-interface designers
	S7L5	List and shortly describe the guidelines for user interface design;
	S7L6	List and shortly describe the principles for user interface design

Table 2. PSLE components for instructional design adapted by [19]

PSLE component	Description
Authentic problems	Problems should be set within context with regard to the students. Students tend to be more engaged on an intellectual level in the learning process when the problem is relevant to them
Information Processing	Computational concepts are acquired through information processing techniques that focus on mental model constructions (constructivism)

(*continued*)

Table 2. (*continued*)

PSLE component	Description
Scaffolding	Scaffolding the program construction into smaller more manageable tasks. This step can potentially develop and foster all three dimensions of computational thinking (concepts, practices and perspectives). This step is based on the constructionism process where meaningful products are built for themselves or others as an end result of continuous knowledge construction
Reflection	Students reflect on computational processes and their programming process, either by self-reflection or peer-reflection. Reflection allows the student time to think about his/her own performance and potentially realising his/her misconceptions and shortcomings

Table 3. Concept videos.

Concept Video #	Title	Duration (minutes)	Learning Outcome (Table 2)
CV1	Introduction to Visual Studio	23:55	S1L1, S1L2, S1L3, S1L4, S1L5, S1L6, S1L7, S1L8, S1L9, S1L10
CV2	TextBox, Variables and Calculations	25:02	S2L1, S2L2, S2L3, S2L4
CV3	Exception Handling	10:45	S2L5
CV4	Constants, Field Variables, Math Class, Debugger	19:00	S2L6, S2L7, S2L8, S2L9, S2L10
CV5	Decision Structures	11:44	S3L1
CV6	Equality, Relational and Logical Operators	10:17	S3L2
CV7	Conditional Expressions with Bool Variables and Flags	7:34	S3L3
CV8	Comparing Strings	8:00	S3L4
CV9	Using TryParse	8:57	S3L5
CV10	Radio Buttons, Checkboxes and Listboxes	8:17	S3L6, S3L7
CV11	Switch Statements	10:45	S3L6, S3L7
CV12	More Listbox Methods	7:25	S3L8, S4L1
CV13	While Loops	7:01	S4L2
CV14	Plus Plus and Minus Minus	2:26	S4L3

(*continued*)

Table 3. (*continued*)

Concept Video #	Title	Duration (minutes)	Learning Outcome (Table 2)
CV15	Do-While Loops	5:45	S4L5
CV16	For loops	9:23	S4L4
CV17	Random Numbers	5:33	S4L8
CV18	Load Event	3:24	S4L9
CV19	While Loops with Debugger	3:18	S4L2
CV20	Do-While with Debugger	10:05	S4L5
CV21	Writing Textfiles	6:18	S4L6, S4L7
CV22	Reading Textfiles	7:16	S4L6, S4L7
CV23	File Dialogs	7:16	S4L6, S4L7
CV24	Intro to Methods	7:25	S5L1
CV25	Void Methods	6:57	S5L2
CV26	Passing Arguments to Methods	8:24	S5L2
CV27	Passing Arguments by Reference	7:01	S5L2
CV28	Value Returning methods	6:42	S5L3
CV29	Debugging Methods	7:44	S5L4
CV30	Classes and Multiple Forms	12:34	S6L1, S6L2
CV31	Passing Data Between Multiple Forms	6:42	S6L3

3.3 Data Collection, Instrument and Analysis

The Unified Theory of Acceptance and Use of Technology (UTAUT) was formulated by Venkatesh, Morris, Davis, and Davis (2003) by reviewing and synthesizing eight theories/models of technology use. We used UTAUT as a lens to make sense of and obtain a deeper understanding of the factors determining the success of online learning videos within a programming course. The UTAUT was developed with the following underlying factors: performance expectancy, effort expectancy, social influence, facilitating conditions, self-efficacy, anxiety, attitude towards using technology and behavioural intention. The definitions of the factors are given in Table 4.

A questionnaire regarding the concept videos on the LMS was developed and the students entered their biographic data in the first section. Venkatesh, Morris, Davis and Davis [20]) and Hardgrave, Davis and Riemenschneider [8]) constructed questionnaires based on measurement scales from previous research and the factors considered in this

Table 4. The UTAUT model factors and definitions adapted by [10]

Factor	Definition
Performance expectancy	The degree to which a person believes that using the system will help him/her to better his/her performance and therefore enhance the quality of his/her work (Venkatesh et al., 2003)
Effort expectancy	The degree of ease that is associated with the use of a certain system (Venkatesh et al., 2003)
Social influence	Refers to the degree to which a person experiences interpersonal influence to use a system from important people within his/her social environment (Venkatesh & Davis, 2000)
Facilitating conditions	"the degree to which an individual believes that an organizational and technical infrastructure exists to support the use of the system" (Venkatesh et al., 2003)
Self-efficacy	A person's belief in his/her own ability to succeed in a specific situation or in accomplishing a task (Bandura, 1995)
Anxiety	A sense of worry, nervousness, or unease about something with an uncertain outcome (Oxford English Dictionary, 2014)
Attitude towards using technology	A person's overall affective reaction to using a system (Venkatesh et al., 2003)
Behavioural intention	An individual's intention to use an innovation in the future, whether or not he or she used it currently (Ajzen, 1991)

study are: Performance expectancy (PEx), Effort expectancy (EfEx), Attitude towards using technology (Att), Facilitating conditions (FC) and Behavioural intention (BI). The factors of social influence, self-efficacy and anxiety were not considered applicable to answering the research question and were therefore excluded.

In the second section of the questionnaire, 21 questions were developed based on the UTAUT factors (see Addendum A). The questions were accompanied by a five-point Likert response scale from 1 (Strongly disagree) to 5 (Strongly agree). Lastly, two open-ended questions were asked: "*Please list any benefits of the concept videos*" and "*Please list any drawbacks of the concept videos*".

Students taking the course received a link to the anonymous online questionnaire via the e-learning system after the final assessment, but before the final marks were released. A total of 509 usable responses were received, representing an overall response rate of 80.5%.

The 509 responses were examined using the five variables. A Cronbach's α coefficient was calculated for each of the five factors and it was found (as shown in Table 5) to be reliable ($\alpha \geq 0.60$) for all five factors.

Table 5. Factors with Reliability Coefficients

Factor	Number of items	Cronbach's alpha (α)
Performance expectancy (PEx)	5	0.91
Attitude towards using technology (ATT)	4	0.90
Effort expectancy (EfEx)	5	0.89
Facilitating conditions (FC)	5	0.86
Behavioural intention (BI)	2	0.72

For the analysis of quantitative data, SPSS Version 26 was used, and for the analysis of qualitative data, Atlas.ti 9.

4 Results and Discussion

4.1 Quantitative Results

The statistical results are summarized in Table 6 and show that the mean values of all five factors are relatively high.

FC with the highest mean showed that students believe that the concept videos fit well with the way they study and that they have the resources necessary to support the use of the concept videos. It is a significant finding in view of the challenges faced by South-African students in terms of access to devices and data during the Covid-19 lockdown and emergency online teaching and learning. In addition, BI showed students' willingness to use concept videos in other modules and in their future studies if they are provided.

PEx showed that the students believe that the concept videos are useful and can improve their performance by saving time and increasing their productivity. This is significant since Fidalgo, Thormann, Kulyk and Lencastre [7] found that one of the students' major concerns about online learning was time management.

With Att, the students indicated that using the concept videos was enjoyable and made the course more interesting. EfEx also has a high mean indicating students believe that using the concept videos is effortless and uncomplicated.

It is therefore clear from Table 6 that the students in this study had a positive experience with using the concept videos in the programming course with most factors showing a relatively high mean.

4.2 Qualitative Results – "Benefits"

Themes identified in the qualitative data regarding the benefits of concept videos included the following:

Ease of use and repeatability.
Many respondents found the videos easy to use and understand. It was reported as practical, easy to locate, use, and rewatch the videos as needed.

Table 6. Descriptive Statistics (n = 509)

Factor	Mean*	Std. Deviation
Facilitating conditions (FC)	4.18	0.68
Behavioural intention (BI)	4.12	0.83
Performance expectancy (PEx)	3.96	0.84
Attitude towards using technology (Att)	3.87	0.84
Effort expectancy (EfEx)	3.84	0.81

* Likert-style responses were ranked from 1 to 5 respectively

The concept videos are easy to use and during one's personal revision, they can easily go back to rewatch the video regarding the topic area they did not fully grasp

Short videos.
It was a common theme amongst participants that they liked the fact that the videos are short and focused. This concurs with the findings of [5] and [13].

The shorter length of the concept videos compared to other subjects make them easier to use and to understand.

The concept videos are short, but effective. Only necessary information was talked about therefore it held my attention effectively. The concept videos made use of easy understandable language which is helpful to someone who has never coded before.

Relevance and conciseness.
Many participants reported that the videos were useful to help them understand the specific concept being taught and thus help them complete practical tasks.

By dividing the videos into different focus points helps orientate me and successfully helps me to understand the practical's better

Stimulates your brain more when it comes to studying because slides aren't always effective. And they help you understand how to apply the information when you code

Promote understanding
The thoroughness and clarity of explanations helped most participants to build and enhance their mental models about the concepts which confirms the principle set by Mayer [14] of managing essential overload.

Using the concept videos is extremely helpful in understanding how to apply the theoretical concepts practically.

I was able to understand the things I did not understand when using the e-book.

Learning and understanding new concepts are easier

Engagement and presentation.
Many participants commented positively on the presenter's delivery and how the videos were perceived as fun and friendly.

They were thorough enough to explain how the concepts work, but they didn't spoon feed you (you had to think further - which I liked because it was a challenge and rewarding to figure it out). Furthermore, they were friendly and fun and the student presented them VERY well.

Make learning and understanding new concepts easier and fun

It made studying interesting

Promote efficiency.
Several participants mentioned that they could utilize their time more efficiently when watching the videos since they understood the work better and could complete practical work faster.

They helped me work on an amount of time allocated for the particular work since I don't have to do so much research on the work

The time used to watch videos saved me a lot of time when doing the practical's

4.3 Qualitative Results – "Drawbacks"

The qualitative data revealed considerably more advantages in respect of the concept videos than difficulties or frustrations. Themes that emerged as drawbacks include:

Lack of videos on complex problems.
Many students mentioned that they want more videos solving complex problems. The aim initially was to explain a concept in a way that will build mental models. Longer online workshops that covered solving complex problems were also held. Students need to apply concepts learned. In our view, this theme is a win since we want students to think and apply concepts learnt.

Some of the videos were only the basics about the topic and in the practicals you had to think a bit harder

Additional knowledge may be left out and it does not detail every aspect of the tasks

No interaction
Less than ten students indicated that they could not ask questions since it was a video.

I sometimes had questions, but this drawback was largely mitigated by the weekly tutorials and the ease of access to my demi for questioning

Time-consuming
Some students (only about 10) felt that the videos were time-consuming.

No drawbacks
Many students indicated that the videos had no drawbacks at all.

There are none

Not any I have seen yet

5   Conclusion

The majority of participants reported an improvement in academic achievement due to the use of the videos, and responses suggest that participants considered concept videos to be very helpful for learning programming concepts. They liked how simple, concise, and straightforward the concept videos were. The visual aspect of the video that shows how to code and implement concepts made the content more understandable. The concept videos also brought a level of convenience since students could go back for reference. Other benefits reported include:

- Increased efficiency in task accomplishment and work quality through the use of the videos.
- Low effort including mental effort required to use the videos.
- Enjoyment and interest in the course through the use of the videos.
- Compatibility of the videos with the students' learning styles.

The key factors of the success of online learning concept videos for programming are:

- Ease of use and repeatability.
  Keeping a consistent learning management system layout for every week and posting the videos using the same layout and structure.
- Reasonably short duration.
  It was commented repeatably that the duration of the videos was preferable.
- Relevance and conciseness.
  Aligning the videos with the outcomes is key. Keep the videos focused on the concept.
- Thoroughness of videos.
  Include all relevant information without giving too much information. Try to keep with a concept a video.
- Engagement and presentation.
  A well-prepared presenter enhances the experience.
- Resources
  Availability of necessary resources to use the videos.

- Knowledge
  Ensure that students have the necessary knowledge to use the course's videos. We made a compulsory orientation section and video which explained the course layout, tools and expectations.

Drawbacks reported include not having assistance and interaction available when technical and content-related difficulties occur with the videos. Addressing these drawbacks would enhance the overall experience of using concept videos.

There have been many studies on online learning videos for general content, but this study offers insights for lecturers of specifically programming courses to create effective online learning videos. It is recommended that lecturers of programming courses should consider student feedback when designing online courses in order to create engaging and effective online videos to establish best practices in online teaching and learning.

Addendum A

Factors	Item code	Item
Performance expectancy (PEx)	PEx1	The concept videos were useful in my studies
	PEx2	The quality of my work is enhanced by the use of the concept videos
	PEx3	Using the concept videos improved my academic achievement
	PEx4	The advantages of using the concept videos outweigh the disadvantages
	PEx5	Using the concept videos enables me to accomplish tasks more quickly
Effort Expectancy (EfEx)	EfEx1	Using the concept videos was easy for me
	EfEx2	Using the concept videos does not require a lot of mental effort
	EfEx3	I find the concept videos easy to use
	EfEx4	I think the concept videos are clear and understandable
	EfEx5	Using the concept videos does not require a lot of effort
Attitude towards using technology (Att)	Att1	I like working with concept videos
	Att2	The concept videos make the course more interesting
	Att3	Working with the concept videos is fun

(*continued*)

(*continued*)

Factors	Item code	Item
	Att4	Using the concept videos is a good idea
Facilitating conditions (FC)	FC1	Using the concept videos fits well with the way I study
	FC2	I have the resources necessary to use the eFundi site's concept videos
	FC3	I have the knowledge necessary to use the course's concept videos
	FC4	The concept videos are compatible with the way I study
	FC5	Assistance is available when difficulties with the concept videos occur
Behavioural intention (BI)	BI1	Given the opportunity, I would use the concept videos in my studies
	BI2	If concept videos were available for other modules, I predict I would use them

References

1. Ayebi-Arthur, K.: E-learning, resilience and change in higher education: Helping a university cope after a natural disaster. E-learning and Digital Media **14**, 259–274 (2017)
2. Bayne, S., Ross, J.: 'Digital Native' and 'Digital Immigrant' Discourses. In: Land, R., Bayne, S. (eds.) Digital difference: Perspectives on online learning, pp. 159–169. Sense Publishers, Rotterdam (2011)
3. Becker, S.A., Cummins, M., Davis, A., Freeman, A., Hall, C.G., Ananthanarayanan, V.: NMC horizon report: 2017 higher education edition. The New Media Consortium (2017)
4. Choe, R.C., et al.: Student satisfaction and learning outcomes in asynchronous online lecture videos. CBE—Life Sciences Education **18**, ar55 (2019)
5. Diwanji, P., Simon, B.P., Märki, M., Korkut, S., Dornberger, R.: Success factors of online learning videos. In: 2014 International Conference on Interactive Mobile Communication Technologies and Learning (IMCL2014), pp. 125–132. IEEE (2014)
6. Elçiçek, M., Karal, H.: A framework proposal for the design of video-assisted online learning environments for programming teaching. Ilkogretim Online **19**, 1820–1837 (2020)
7. Fidalgo, P., Thormann, J., Kulyk, O., Lencastre, J.A.: Students' perceptions on distance education: a multinational study. Int. J. Edu. Technol. High. Edu. **17** (2020)
8. Hardgrave, B.C., Davis, F.D., Riemenschneider, C.K.: Investigating determinants of software developers' intentions to follow methodologies. J. Manag. Inf. Syst. **20**, 123–151 (2003)
9. Johnson, L., Adams, S., Cummins, M., Estrada, V., Freeman, A., Hall, C.: NMC horizon report: 2016 higher education edition. The New Media Consortium, Austin, TX (2016)

10. Liebenberg, J., Benadé, T.: Learning styles and technology acceptance. In: Proceedings of the 9th Annual ISTE Conference on Mathematics, Science and Technology Education, pp. 353–361. UNISA (2018)

11. Liebenberg, J., Benadé, T., Ellis, S.: Acceptance of ICT: Applicability of the Unified Theory of Acceptance and Use of Technology (UTAUT) to South African Students. The African J. Info. Sys. **10**, 1 (2018)

12. Lye, S.Y., Koh, J.H.L.: Review on teaching and learning of computational thinking through programming: What is next for K-12? Comput. Hum. Behav. **41**, 51–61 (2014)

13. Manasrah, A., Masoud, M., Jaradat, Y.: Short videos, or long videos? a study on the ideal video length in online learning. In: 2021 international conference on information technology (ICIT), pp. 366–370. IEEE (2021)

14. Mayer, R.E.: The Cambridge Handbook of Multimedia Learning. Cambridge University Press (2014)

15. Pandey, A.: 5 Amazing eLearning video strategies to keep your digital learners hooked. eLearning Industry (2020)

16. Parker, A., van Belle, J.-P.: The iGeneration as Students: Exploring the Relative Access, Use of, and Perceptions of IT in Higher Education. In: Liebenberg, J., Gruner, S. (eds.) CCIS 730, pp. 3–18. Springer (2017)

17. Pelletier, K., et al.: 2022 EDUCAUSE Horizon Report Teaching and Learning Edition. EDUC22 (2022)

18. PNET: Job Market Trends Report Q1 2023 (2023)

19. van der Linde, S., Liebenberg, J.: Utilizing Computational Thinking in Programming to Reduce Academic Dishonesty and Promote Decolonisation. ICT Education. SACLA 2021. Communications in Computer and Information Science **1461**, 51–66 (2022)

20. Venkatesh, V., Morris, M.G., Davis, G.B., Davis, F.D.: User acceptance of Information Technology: Toward a unified view. MIS Q. **27**, 425–478 (2003)

Exploring Flipped Learning in an Introductory Programming Module: A Literature Review

Nita Mennega(⊠) ⓘ and Tendani Mawela ⓘ

Department of Informatics, University of Pretoria, Pretoria, South Africa
{nita.mennega,tendani.mawela}@up.ac.za

Abstract. Introductory programming courses form the foundation of all programming qualifications, which are in turn essential to a well-educated information technology workforce. Sadly these courses cannot boast high retention or success rates. The inverted (or "flipped") classroom approach has been touted as a way to engage large classes of mixed-ability students. To inform the design of a first-year university introductory programming course, the authors conducted a systematic literature review on existing flipped classroom implementations. Five electronic databases were searched with the keywords "flipped learning" and "introductory programming" for peer-reviewed academic literature published in the decade between 2013 and 2022 to answer the question "How may flipped learning be used to enhance student learning in an introductory programming module?". It was found that while the flipped learning approach poses challenges to both lecturers and students, it creates an increase in engagement during class and ensures that better use is made of limited face-to-face time between instructors and students. Meaningful future research would focus on seeking objective student feedback to understand the nature of the student experience.

Keywords: Flipped Classroom · Flipped Learning · H5P · Instructional Videos · Introductory Programming

1 Introduction

Programming is a core skill in the Information Technology (IT) related higher education programmes such as computer science and informatics as well as other science, technology, engineering, and mathematics (STEM) related degrees. The literature indicates that despite the advances in pedagogy and the integration of programming into some high school IT curricula the failure and dropout rates of first year university programming courses continue to be troubling [1, 2].

Teaching and learning how to program is no simple task. Educators have to simultaneously teach computational thinking and programming language syntax while students are easily discouraged by the strangeness and complexity of programming. Traditional teaching approaches allow the instructor to systematically introduce new concepts to students, who are able to converse with their instructor and peers to clarify misconceptions.

© The Author(s), under exclusive license to Springer Nature Switzerland AG 2024
H. E. Van Rensburg et al. (Eds.): SACLA 2023, CCIS 1862, pp. 64–74, 2024.
https://doi.org/10.1007/978-3-031-48536-7_5

Rapidly growing class sizes and the recent pandemic has forced education online, removing the valuable class interaction and limiting communication to emails and face-less online sessions. With the relaxing of social distancing measures and the concomitant move back to in-person classes, educators have a golden opportunity to harness the material they developed during the pandemic, in tandem with face-to-face classes, to provide a richer offering to students, especially those who may be struggling with the topic.

The flipped classroom (FC) is a pedagogical method that represents a unique combination of learning theories previously thought to be incompatible: active learning (founded upon a constructivist ideology) and instructional lectures (based on behaviourist principles), all of which enabled by technology. The interest in harnessing the flipped classroom approach is borne out by a surge of publications regarding the inverted classroom approach [3, 4]. This approach involves well-planned pre-class, in-class and post-class activities, where students are expected to take an active approach in preparing before class and participating during class. The term "flip" indicates a reversal of the traditional teaching model to achieve more effective learning, by presenting students with pre-recorded videos for low-level cognitive learning before class, which prepares for high-level cognitive learning happening in the classroom [5]. The flipped classroom model has at times proven its worth in higher student participation rates and better grades. However, it requires significant educator effort and a sometimes-unrealistic level of maturity from the student.

2 Research Method

In order to build a firm base from which to create a flipped classroom implementation, a systematic literature review was conducted by following the guidelines proposed by [6]. This involved following a highly structured process that includes the following seven steps: specifying a research question, conducting searches of academic databases, selecting studies, filtering the studies according to their relevance, extracting data, synthesising the results, and writing the review report.

2.1 Research Question

The research question for this study was kept as simple as possible:

How may flipped learning be used to enhance student learning in an introductory programming module? The goal is to use the review to determine proven ways to harness the flipped learning approach to present a programming course to students who have never programmed before.

2.2 Search Terms

The search terms were constructed from only two keywords: "Flipped Learning" and "Introductory Programming". Synonyms were included which led to the search string ("Flipped classroom" OR "Flipped Learning") AND ("Introductory Programming" OR "introduction to programming" OR "novice programmers" OR "learn programming" OR "learning to program" OR "teach programming").

Five databases were searched on 22/06/2022 using the individual search terms as described in Table 1. The search was performed for papers published from 2013 to 2022. When any of these papers cite other relevant papers, the cited papers may also be included (under "other sources" – see Fig. 1).

Table 1. Database search strings and results

Database Name	Search term	#
ACM digital library	[[All: "flipped classroom"] OR [All: "flipped learning"]] AND [[All: "introductory programming"] OR [All: "introduction to programming"] OR [All: "novice programmers"] OR [All: "learn programming"] OR [All: "learning to program"] OR [All: "teach programming"]]	164
IEEE Explore	("Flipped classroom" OR "Flipped Learning") AND ("Introductory Programming" OR "introduction to programming" OR "novice programmers" OR "learn programming" OR "learning to program" OR "teach programming")	15
Science Direct	("Flipped classroom" OR "Flipped Learning") AND ("Introductory Programming" OR "introduction to programming" OR "novice programmers" OR "learn programming" OR "learning to program" OR "teach programming")	55
Springer Link	("Flipped classroom" OR "Flipped Learning") AND ("Introductory Programming" OR "introduction to programming" OR "novice programmers" OR "learn programming" OR "learning to program" OR "teach programming")'	229
Scopus	TITLE-ABS-KEY (("Flipped classroom" OR "Flipped Learning") AND ("Introductory Programming" OR "introduction to programming" OR "novice programmers" OR "learn programming" OR "learning to program" OR "teach programming"))	52

- Number of articles found

As can be seen from Table 1, a total of 515 citations were downloaded from the electronic databases.

2.3 Selection Criteria

The papers that were included were those written in the English language and published between 2013 and 2022. Only peer-reviewed journal articles and conference proceedings were selected. Two opinion papers that appeared in industry magazines were also included.

We focused on the flipped learning approach to teach an introductory programming course at tertiary level. For this reason, literature that pertained to secondary education was excluded. Literature that pertained to computational thinking per se was also excluded. The term "inverse class" was omitted from the search terms because its use has been superseded by the terms "flipped learning" or "flipped classroom". Citations where the full-text could not be sourced, were excluded.

2.4   Source Selection

The PRISMA flowchart in Fig. 1 illustrates the screening process by showing the subsequent phases the literature was subjected to.

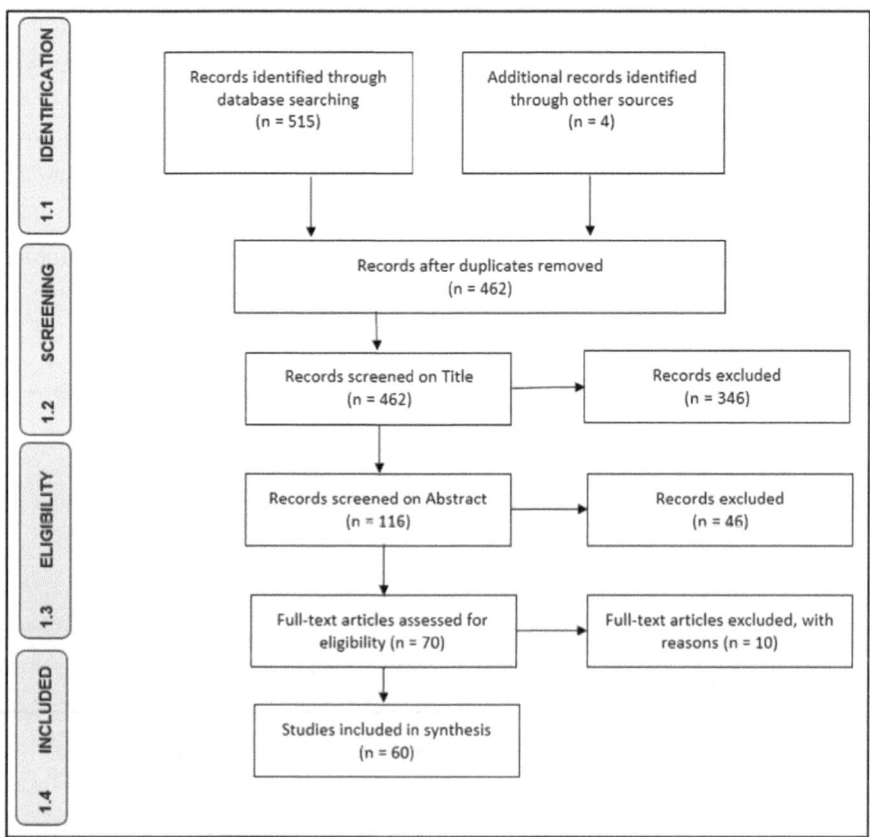

Fig. 1. Systematic Literature Review (SLR) process in the form of a PRISMA Flowchart.

Starting with a total of 519 citations, duplicates were removed before the screening started. The titles and then the abstracts were reviewed, with 346 and 46 papers respectively discarded in these screening phases. Ten papers were found ineligible after perusing their full text, with reasons such as full text inaccessibility and unrelated foci such as discussions of programming environments or computational thinking. This left 60 papers for inclusion in this review.

2.5   Data Extraction

Figure 2 shows the publications per year and the types of publication of the final 60 studies included in the review. It illustrates a decade of growth in publications on the flipped classroom approach. Interest in the topic spiked during 2019.

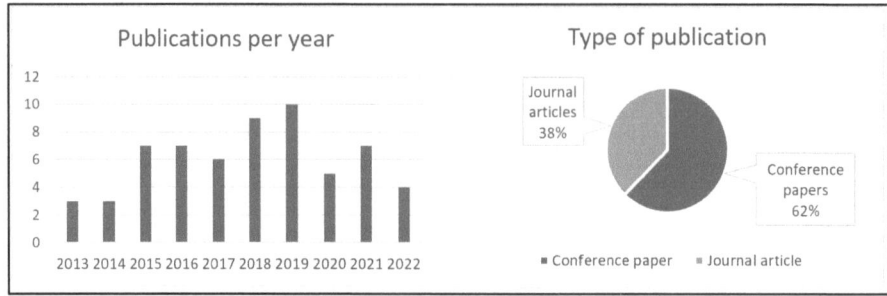

Fig. 2. Publications published yearly and the type of publications included in review.

Most of the research published on the flipped classroom approach, is in the form of case studies (see Fig. 3). Three-quarters (45 out of 60) of the papers under review are devoted to descriptions of existing flipped classroom implementations.

Research Strategy	
Case study	45
Review	7
Discussion / Panel discussion	2
Thematic analysis	1
Quant study - learning analytic data	1
Grounded Theory	1
Action Research	1
Interviews and focus-group discussions	1
Instrument measuring self-efficacy	1

Fig. 3. Research strategies followed in identified literature.

Following descriptions of flipped classroom implementations, a surprisingly large percentage of the studies in this review, are in themselves literature reviews: 11.6% or 7 papers out of a total of 60. This together with the fact that most of the research on flipped classrooms has been performed during the last five years, may indicate that educators worldwide are experiencing similar problems in teaching introductory programming and that many are reviewing the literature before attempting to "test the waters" by implementing the inverted classroom model in their own teaching. The interest in the flipped classroom approach to teach introductory programming is illustrated by two papers that report on panel discussions devoted to the topic. Most of the papers have been presented and discussed at conferences (62%), while 38% of the papers comprise journal articles.

The USA takes the lead in the number of papers published on the topic (21), followed by Canada (6) and the Netherlands (4). The UK, Finland, India and the Middle East immediately follow them with an average of three publications each. Figure 4 illustrates the countries of origin of the publications in this review.

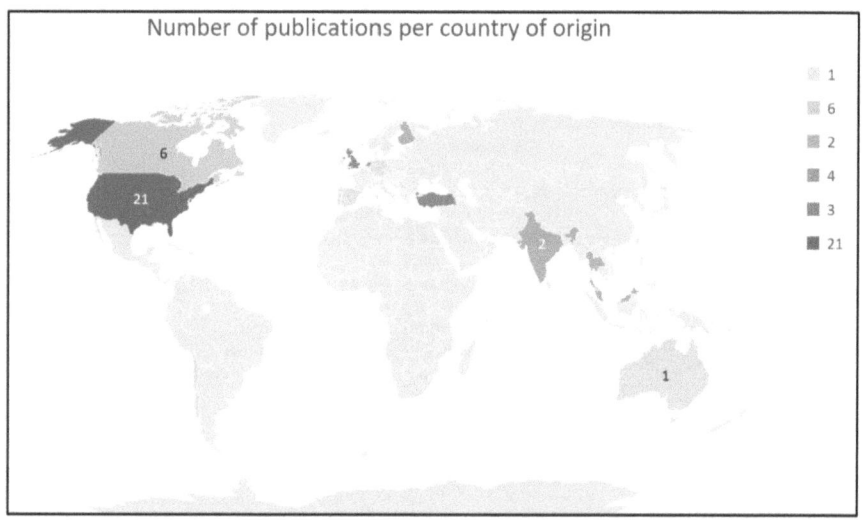

Fig. 4. Number of publications per country of origin

The map shows that there is a wide variety of countries publishing on the topic, with the USA leading on publications, followed by European countries, Asian countries, and Australia.

3 Results

To analyse the data, we employed a thematic analysis approach following the guidelines by Braun and Clarke [7]. Thematic analysis is a flexible technique for methodically identifying, coordinating, and gaining insight into patterns of themes across a data set. It consists of six phases that comprise familiarization during data collection, coding to highlight similarities across the data, generating themes based upon emerging patterns, reviewing and defining these themes and lastly writing up the analysis.

Three main categories emerged from the literature surveyed: lecturers' approaches to teaching via the flipped classroom approach, the students' reaction to the flipped classroom approach, and suggestions for future study. The categories and themes are illustrated in Fig. 5.

The first two categories, namely the approaches to teaching and the reaction of the students to the flipped classroom approach are illustrated in Fig. 6. They consist of positive as well as negative aspects of the flipped classroom approach and include the points of view of both the lecturers and the students.

From the lecturers' point-of view, the main drawback is that preparing videos for pre-class viewing is extremely time consuming [8–11]. Once these have been created, however, the videos are effective in teaching course content, especially when there are interactive questions embedded [12]. These questions ensure that students pay attention to the content of the video as they need to answer their pre-class quiz [4, 13]. Another

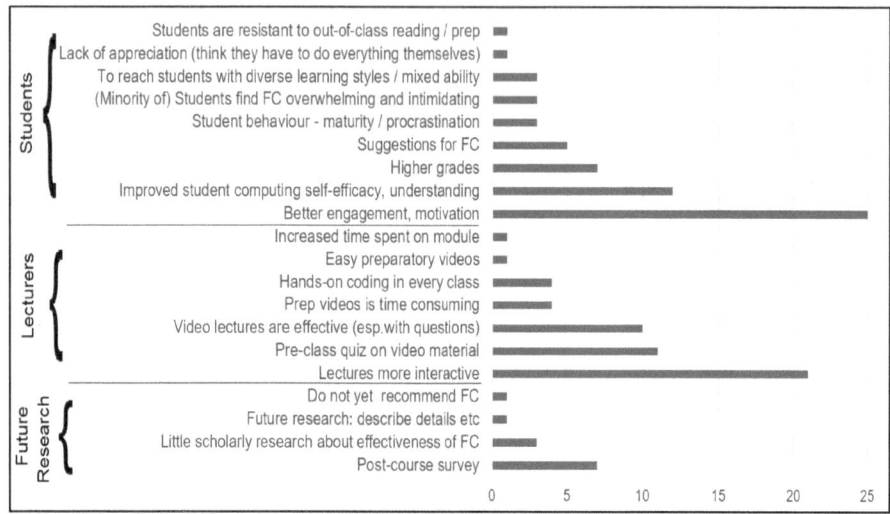

Fig. 5. Themes identified from literature

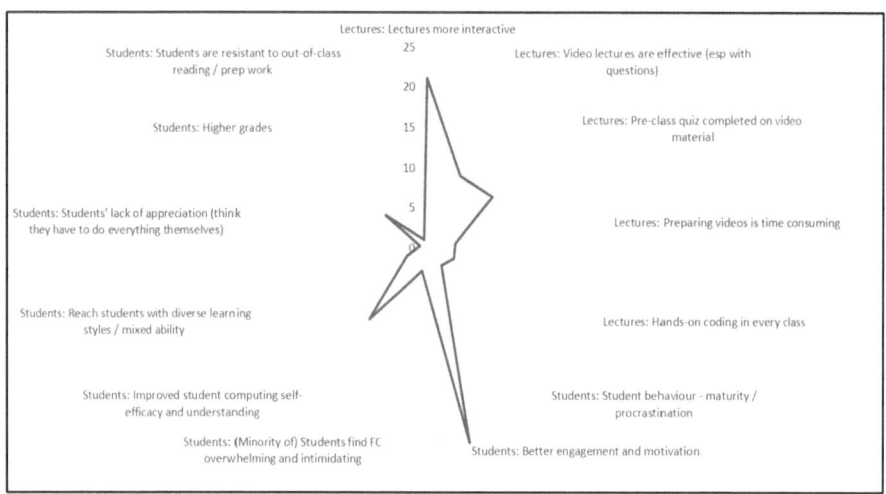

Fig. 6. Radar chart illustrating aspects of flipped classroom implementations

advantage is that the lectures are more interactive [14, 15]. There is also more time for hands-on coding in every class which further improves learning [16, 17].

From the students' point-of view, the flipped classroom approach presents a challenge to their learning behaviour. Many first-year students lack the maturity to work independently and procrastination is a significant obstacle [18]. Some students, albeit the minority, find the flipped classroom approach overwhelming and intimidating [19]. One study found that students are resistant to out-of-class reading and other preparatory work [20] and another study found that students lack appreciation for the approach and think that they are expected to do all the coursework by themselves [10]. However, the majority of studies found that the flipped classroom approach resulted in better engagement and motivation (25 studies in the review). A further twelve studies reported improved student computing self-efficacy and understanding, and seven studies reported an increase in grades. It was also found that the flipped classroom approach excels at reaching students with diverse learning styles and mixed abilities [21, 22].

The third category, suggestions for future study, include calls for future researchers to describe their case studies in precise detail, so that they may be replicated by others. Giannakos [23] call for quantitative research to determine the barriers faced in the flipped classroom approach by the students who battled adapting to it. Two studies profess that little scholarly research has been done regarding the effectiveness of the flipped classroom approach [24, 25]. To assist in determining the effectiveness of particular interventions, Luxton-Reilly et al. [3] call for researchers to identify various tasks that strongly influence the student experience. McCord and Jeldes [21] call for research on designing flipped classroom interventions that enable students to develop self-regulation skills, which would allow them to take more responsibility for their own learning.

4 Discussion

With the increasing urgency of successfully imparting digital skills to students, and in particular basic programming skills, comes the search for effective teaching methods in introductory programming courses. The flipped classroom approach (also known as inverted learning) is an increasingly popular method to teach introductory programming skills. This study employed a systematic literature review method to investigate academic publications on the topic and identify how best to use flipped learning to enhance student learning in an introductory programming course.

The literature shows that the flipped classroom approach places significant demands on both the lecturer and the student. It requires that the lecturer spend many hours creating preparatory material while it requires a high level of self-directed learning by its students. Many reported case studies profess that their flipped classroom approach resulted in higher grades for their students, but this may be due to publication bias and the longer requisite hours being spent on the course material. This study found that only 7 out of 45 case studies reported an outcome of a higher final score for the students of the course. Despite this disappointing result, the flipped classroom approach has been found to consistently contribute to better engagement of students with the subject matter.

Sobral [4] summarised the outcomes of prevailing studies on the flipped classroom approach with some interesting findings, such that students prefer the flipped classroom approach to traditional lectures, because they would rather use their class time for problem-solving with the instructor present, rather than listening passively to a lecture. Students' enthusiasm for the course increased over the semester and they grasped the importance of thoroughly practicing programming.

With the aim of enhancing student learning, lecturers should follow a ternary approach by preparing not only their study material, but also their course structure and their students. To support students who find the pre-class videos intimidating, and to counter resistance to out-of-class reading, lecturers can provide a roadmap of all activities, each carrying a mark value. Jovanovic et al. stresses the crucial role of course design. They found that appropriate assistance in the flipped classroom structure of a module can be achieved by nudging students to complete certain activities, informed by analytics from the institution's learning management system [26].

Preparing lectures is time-consuming but it may prove an ideal opportunity to repurpose existing recordings created for online teaching during the Covid-19 lockdown. Bishop and Verleger [24] showed that video lectures (slightly) outperform face-to-face lectures, but that interactive online videos do even better. An advantage of certain Learning Management Systems is the functionality to create interactive online videos that integrate with the student marks grid, so that the marks attained are automatically recorded as students complete their interactive videos.

Sobral [4] confirms that when lecturers add explicit learning objectives and exercises to be completed before students come to class, student preparedness is enhanced and class-time may be used for advanced learning objectives. She also confirms that that the flipped classroom approach improves competency acquisition.

The fact that the flipped classroom approach consistently contributes to better student engagement, a fruitful avenue of future research would be to focus on seeking objective student feedback to understand the nature of the student experience and why the improved engagement does not translate into higher student grades, with the aim to close the loop for the students to reach their goal of mastering the course material and gaining better marks.

5 Conclusion

This study identified two categories of issues to consider when implementing an effective flipped learning approach to enhance students' learning of the concepts in introductory programming. The two categories pertained to lecturers' concerns (teaching material and course structure) and managing learning problems (students' reaction to the flipped classroom approach). The consensus is that the flipped classroom approach contributes to better use of limited face-to-face time between instructors and students, but that it requires significant effort and dedication from both teachers and students. A significant advantage offered by the flipped learning approach is its proven success in teaching mixed-ability introductory programming courses [15, 22].

A third category pertained to future research on the topic. Luxton-Reilly et al. summarises future research in their excellent review on the topic of introductory programming by saying "The perception, and ultimately the success, of computing as a discipline could be substantially improved by continued reporting of case studies and experiences of engaging activities that teachers could adopt in their classrooms, and by increased dissemination encouraging the adoption of these approaches." [3].

References

1. Medeiros, R.P., Ramalho, G.L., Falcao, T.P.: A systematic literature review on teaching and learning introductory programming in higher education. IEEE Trans. Educ. **62**, 77–90 (2019). https://doi.org/10.1109/TE.2018.2864133

2. Vihavainen, A., Airaksinen, J., Watson, C.: A systematic review of approaches for teaching introductory programming and their influence on success. In: ICER 2014 – Proceedings 10th Annual International Conference International Computing Education Research, pp. 19–26 (2014). https://doi.org/10.1145/2632320.2632349

3. Luxton-Reilly, A., et al.: Introductory programming: a systematic literature review. In: Annual Conference on Innovation and Technology in Computer Science Education, ITiCSE, pp. 55–106. Association for Computing Machinery (2018)

4. Sobral, S.R.: Flipped classrooms for introductory computer programming courses. Int. J. Inf. Educ. Technol. **11**, 178–183 (2021)

5. Kuo, Y.C., Lin, Y.H., Wang, T.H., Lin, H.C.K., Chen, J.I., Huang, Y.M.: Student learning effect using flipped classroom with WPSA learning mode – an example of programming design course. Innov. Educ. Teach. Int. **60**, 1–12 (2022). https://doi.org/10.1080/14703297.2022.2086150

6. Kitchenham, B., Charters, S.: Performing systematic literature reviews in software engineering. In: Proceeding 28th International Conference Software Engineering – ICSE '06. 2, 1051 (2007). https://doi.org/10.1145/1134285.1134500

7. Braun, V., Clarke, V.: Thematic analysis. APA Handbook of Research Methods in Psychology, Vol 2 Research Designs Quantitative Qualitative Neuropsychological Biological 2, 57–71 (2012). https://doi.org/10.1037/13620-004

8. Vasilchenko, A., Venn-Wycherley, M., Dyson, M., Abegão, F.R.: Engaging science and engineering students in computing education through learner-created videos and physical computing tools. In: ACM International Conference Proceeding Series. Association for Computing Machinery (2020)

9. Erdogmus, H., Péraire, C.: Flipping a graduate-level software engineering foundations course. In: Proceedings of the 39th International Conference on Software Engineering: Software Engineering and Education Track (ICSE-SEET '17) (2017)

10. Heines, J.M., Popyack, J.L., Morrison, B., Lockwood, K., Baldwin, D.: Panel on flipped classrooms. In: SIGCSE 2015 – Proceedings of the 46th ACM Technical Symposium on Computer Science Education, pp. 174–175. Association for Computing Machinery (2015)

11. Maher, M.L.M., Latulipe, C., Lipford, H., Rorrer, A.: Flipped classroom strategies for CS education. In: SIGCSE'15, March 4–7, 2015, pp. 726. Kansas City, MO, USA. ACM (2015)

12. Shaarani, A.S., Bakar, N.: A new flipped learning engagement model to teach programming course. (IJACSA) Int. J. Adv. Comput. Sci. Appl. **12**, 9 (2021)

13. Eusoff, R., et al.: Implementing flipped classroom strategy in learning programming. (IJACSA) Int. J. Adv. Comp. Sci. Appl. **12**(10) (2021). Science and Information Organization (2021)

14. Ruiz de Miras, J., Balsas-Almagro, J.R., García-Fernández, Á.L.: Using flipped classroom and peer instruction methodologies to improve introductory computer programming courses. Comput. Appl. Eng. Educ. **30**, 133–145 (2022)

15. Mohamed, A.: Teaching highly mixed-ability CS1 classes: a proposed approach. Educ. Inf. Technol. **27**, 1–18 (2021). https://doi.org/10.1007/s10639-021-10546-8

16. Gordon, N., Brayshaw, M., Grey, S.: A flexible approach to introductory programming engaging and motivating students. In: ACM International Conference Proceeding Series. Association for Computing Machinery (2019)

17. Pattanaphanchai, J.: An investigation of students' learning achievement and perception using flipped classroom in an introductory programming course: a case study of Thailand higher education. J. Univ. Teach. Learn. Pract. **16**, 36–53 (2019). https://doi.org/10.53761/1.16.5.4

18. Harvey, L., Aggarwal, A.: Exploring the effect of quiz and homework submission times on students' performance in an introductory programming course in a flipped class-room environment. In: 2021 ASEE Annual Conference (2021)

19. Ahmed, B., Aljaani, A., Yousuf, M.I.: Flipping introductory engineering design courses: evaluating their effectiveness. In: IEEE Global Engineering Education Conference, EDUCON, pp. 234–239. IEEE Computer Society (2016)

20. Baldwin, D.: Can we "flip" non-major programming courses yet? In: SIGCSE 2015 – Proceedings of the 46th ACM Technical Symposium on Computer Science Education, pp. 563–568. Association for Computing Machinery (2015)

21. McCord, R., Jeldes, I.: Engaging non-majors in MATLAB programming through a flipped classroom approach. Comput. Sci. Educ. **29**, 313–334 (2019). https://doi.org/10.1080/089 93408.2019.1599645

22. Mohamed, A.: Designing a CS1 programming course for a mixed-ability class. In: Proceedings of the 24th Western Canadian Conference on Computing Education, WCCCE 2019. Association for Computing Machinery, Inc (2019)

23. Giannakos, M.N., Krogstie, J., Chrisochoides, N.: Reviewing the flipped classroom research: Reflections for computer science education. In: Proceedings – CSERC 2014: Computer Science Education Research Conference, pp. 23–29. Association for Computing Machinery (2014)

24. Bishop, J., Verleger, M.: The flipped classroom: a survey of the research. In: 120th ASEE AnnualConference & Exposition (2013)

25. Thongmak, M.: Flipping MIS classroom by peers: Gateway to student's engagement intention. In: 26th International World Wide Web Conference 2017, WWW 2017 Companion, pp. 387–396. International World Wide Web Conferences Steering Committee (2017)

26. Jovanovic, J., Mirriahi, N., Gašević, D., Dawson, S., Pardo, A.: Predictive power of regularity of pre-class activities in a flipped classroom. Comput. Educ. **134**, 156–168 (2019). https://doi.org/10.1016/j.compedu.2019.02.011

Transitioning an Introductory Programming Course into a Blended Learning Format

Aslam Safla$^{(\boxtimes)}$, Hussein Suleman, and James Gain

Department of Computer Science, University of Cape Town, Cape Town, Republic of South Africa

`aslam@cs.uct.ac.za`

Abstract. Teaching Introductory Programming is one of the foundations of Computer Science education and is generally the first course novice students take. It is important for the teachers to motivate students who are learning to program with the help of a variety of teaching methods. Over the years teachers have discussed how to best teach introductory programming, what learning styles and tools to use, how to motivate students and what programming languages should be taught. The steady growth in students' numbers has also contributed to the challenges of teaching this course. Blended learning is one approach to address these challenges. Many studies have concluded that blended learning can be more effective than traditional teaching and can improve the students' learning experience. This paper describes the transitioning of an introductory programming course, as part of a large-enrolment first-year subject, into a blended learning format. Periodic face-to-face lessons and traditional forms of assessment were combined with an integrated learning environment for engaging with video, code and quizzes. The result was a lower staff workload and a learning experience that students were satisfied with. Notably, feedback from students shows that they adopted different tools based on their individual learning preferences.

Keywords: Introductory programming · blended learning · e-learning

1 Introduction

The Introductory Programming courses at University of Cape Town have grown from having about 380 students in 2009 to about 1140 who registered in 2022. This increase in student numbers requires an increase in staff as well as lecture venues. As programming becomes a core skill in many disciplines, it is expected that far more students will need to take these courses in future. It has also been shown that traditional teaching approaches are not adequate in helping students to overcome their learning difficulties [1–3].

E-learning has been successfully introduced to improve the traditional teaching approach [4–6]. With E-learning, students are able to access online pedagogical content at any time and from anywhere outside of the classroom. One form of E-learning is blended learning, which provides an educational experience that combines face-to-face teaching with internet-based technologies to get the best advantages of both these learning

© The Author(s), under exclusive license to Springer Nature Switzerland AG 2024
H. E. Van Rensburg et al. (Eds.): SACLA 2023, CCIS 1862, pp. 75–89, 2024.
https://doi.org/10.1007/978-3-031-48536-7_6

approaches [7]. Blended learning has been shown to provide better quality of learning outcomes [8–10]. For this reason, a number of teachers have attempted to use blended learning in introductory programming courses [11, 12].

With the support of the Centre for Innovation in Learning and Teaching at the University of Cape Town, the School of IT undertook a project for the transition of the Introductory Programming undergraduate courses into a blended learning format. This paper will highlight the steps taken in developing the course, as well as the tools created to provide the students with the best possible learning experience. The course was offered for the first time in semester 1 of 2022. Students' performance and satisfaction were measured.

This paper is organised as follows. Section 2 provides some background information on the use of technology in teaching. Section 3 describes the design of the blended learning introductory programming course. The results are analysed and discussed in Sect. 4. Finally, a conclusion of the paper is provided in Sect. 5.

2 Related Work

The teaching of an introductory programming course is not considered an easy task. Various researchers [13–16] have questioned whether introductory programming courses are achieving the expected results. Studies have shown that introductory programming students need to be motivated to study programming, and that these students are at risk of dropping out or failing introductory programming courses [17]. Traditional teaching approaches do not seem adequate to increase students' motivation [3]. Instead of lectures, educators are encouraged to embrace new teaching methods [18].

Blended learning is defined as "the thoughtful integration of classroom face-to-face learning experiences with online learning experiences" [7]. Common methods used are online and face-to-face lectures, self-paced activities, and online discussion groups. Increasingly, research shows that blended learning can enhance the student learning experience and overcome the shortcomings of traditional teaching approaches [2, 8–10, 19]. According to Lopez-Perez, et al. [20], blended learning produces a high level of effectiveness, motivation, and satisfaction, which creates a positive attitude towards learning in students. One of the biggest advantages of blended learning, compared to the traditional face-to-face method, is that one can have unlimited access to information on the internet.

Alammary, et al. [21] identified five different blended learning components: face-to-face instructor-led, face-to-face collaboration, online instructor-led, online collaboration and online self-paced. Griffin, et al. indicated four pedagogical benefits of online self-paced: (i) allowing students to choose the time most appropriate for their learning; (ii) allowing them to learn at their own desired speed; (iii) providing them with the flexibility to learn in any location; and (iv) allowing them to choose the most appropriate learning strategy [22].

Vo et al. [23] showed that blended learning is very effective in the teaching of Science, Technology, Engineering and Mathematics (STEM) disciplines. There have been recent studies on the use of blended learning in teaching introductory programming [24–27]. Two important factors to consider when designing a blended learning course are the tools used that would provide the learners with features such as code debugging, evaluation and similarity checkers [28], and the approach used in planning the blended learning course [29].

3   Course Design

3.1   Background

Various Introductory Programming (CSC1) courses have been taught at the University of Cape Town for over 50 years. The courses used the traditional method of teaching to introduce the basic concepts of programming, using Python as the introductory programming language, in 48 face-to-face lectures and 12 2-h practical sessions. The course is compulsory for all Computer Science students, as well as some students from the Engineering and Commerce Faculties. Since 2009, student numbers have been increasing steadily, from 380 students registered in 2009 to over 1200 students registered at the start of 2022 (see Fig. 1).

Fig. 1. Number of students registered for CSC1 from 2009 to 2022

The teaching plan for these courses is shown in Table 1. The performance of the students is assessed via weekly programming assignments, 3 practical tests, 2 theory tests and a final examination. The weekly assignments and practical tests are automatically graded using an in-house Automarker system, while the theory tests and examination are paper based. Due to class enrolment sizes, the number of lectures per day went from one lecture per day in 2009, to two (repeated) lectures per day in 2011, and steadily to five (repeated) lectures per day in 2020. Practical sessions also increased from two 2-h sessions per week to twelve 2-h sessions per week. With both lecture room space and lab space becoming more difficult to acquire, and with more interest being shown by more departments to make the Introductory Programming courses mandatory for their degree offerings, the need for an alternate teaching model became even more apparent. To overcome these challenges, the School of IT decided to pursue the option of transitioning this course into a blended learning format. In 2018, the School of IT applied to the Centre for Innovation in Learning and Teaching at the University of Cape Town for assistance in developing this course. Funding was requested for some staff time and some assistance with video editing and course material development. This application was approved in 2019.

Table 1. Teaching Plan for the CSC1 courses

Week:	Topic
1	Introduction to Computer Science: What is Computer Science, Applications of Computing, History of Computing, Computer Hardware (Machine Architecture), Computer Software (System Software, Applications), Algorithms, Programming Languages
2	Introduction to Python Syntax: Basic syntax, variables, operators, comments, expressions, output
3	Conditionals: Boolean expressions and logical conditions, If statements, nested ifs, if-else, if ladders
4	Loops: for, while, nested loops
5	Strings and Input
6	Functions: parameters, return values
7	Testing: debugging, equivalence classes
8	Arrays: lists, dictionaries, sets, multi-dimensional arrays
9	Recursion
10	Sorting and Searching
11	File I/O: text files, exceptions
12	Number Systems: Machine representations of data, Binary operations, Boolean algebra

3.2 Course Development and Implementation

Developing an efficient course relies on sound and established principles and mechanisms, with constant changes to include advanced technologies [30]. In developing the blended Introductory Programming course, the following factors needed some extra consideration: components of blended learning used, online tools created, and assessment formats. These factors will be discussed in the next sections.

3.2.1 Components of Blended Learning Used

The blended Introductory Programming course incorporated online lecture material and face-to-face instructor led lessons. The core course material was provided in the form of daily lecture videos, while students would still attend a single face-to-face lecture per week, where students had an opportunity to interact with teachers on the topics covered in the online material. A weekly online quiz was added to the start of each practical session, allowing for teachers to track the progress of students. The weekly assignments were retained, as were the three practical tests.

3.2.2 Online Tools Created

New technology provides new ways of teaching, and one such way is the ability to give students immediate feedback to their answers. This is a key benefit gained from utilizing educational technology [31]. Immediate feedback has also been found to improve learning results in students [32]. Immediate feedback can, in the best case, provide students with a cognitive conflict between what they thought was correct and what actually is.

Since programming is considered a mental skill that requires lots of practical exercises, an efficient blended learning of introductory programming will require teachers to set up the online materials and all necessary practical exercises for evaluation. To facilitate this process, the Tashiv tool (see Figs. 2, 3 and 4) was developed. The tool allows teachers to upload lecture videos onto the tool, while also adding inline quizzes and coding exercises. While students play the lecture video, students are prompted to complete either a quiz or a coding exercise. The quizzes and coding exercises were set up to provide the students with immediate feedback.

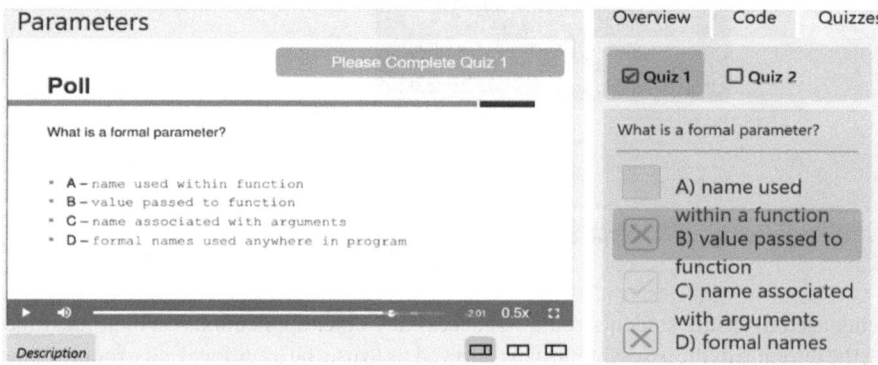

Fig. 2. The Tashiv tool (student view of Quiz)

3.2.3 Assessment Formats

Blended learning combines traditional face-to-face learning and online learning. This helps teachers evaluate the performance of students using two types of assessments [33]: 1) onsite assessment, which is related to the traditional section of blended learning; and 2) online assessment, which is related to the online section of blended learning. While the traditional summative theory tests and examination were retained, both these assessments were moved onto the online platform.

Given the large number of students and the limited lab space, students were allowed to use their own laptops for the online assessments. The assessments were conducted on the Learning Management System (LMS), with the Respondus LockDown browser[1]

---

[1] https://web.respondus.com/he/lockdownbrowser/.

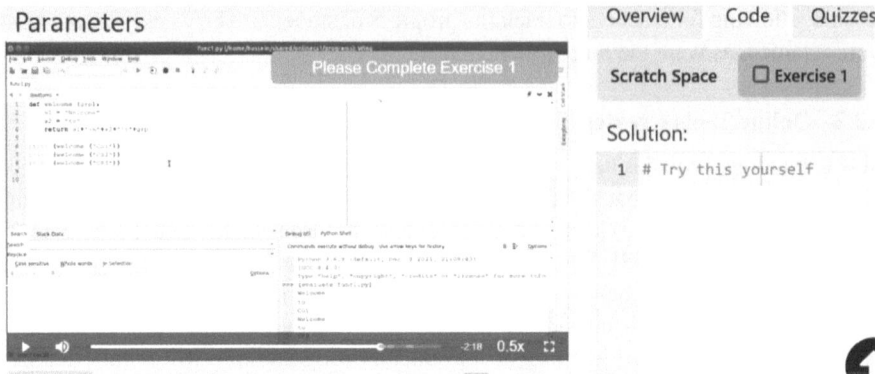

Fig. 3. The Tashiv tool (student view of Coding Exercise)

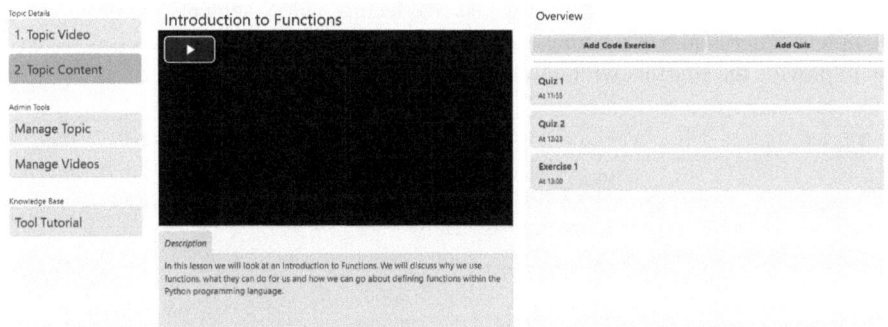

Fig. 4. The Tashiv tool (Admin view)

being used to ensure students could not access any other applications on their machines for the duration of the assessment. This allowed us to use the traditional classroom venues for the assessments. All assessments were invigilated, with a password being shared at each venue at the start of the assessment. One issue that came up in the planning of these assessments was that students might have unreliable computers to take the assessment. A few computer labs at the University were booked for the time of the assessment and students who had unreliable laptops were allowed to take the assessment on the lab computers. Each venue was also supplied with extra plug points for students who needed to charge their laptops during the assessment.

3.3 Implementation

The blended CSC1015F course was first offered in 2022. 1044 students enrolled for the course. The course site was hosted on the LMS (see Fig. 5). Online lecture material was arranged according to sections, with each section having 3 or 4 lessons. Each lesson contained between one and 3 videos, and the set of PowerPoint slides that were used in the lecture videos. As mentioned, some videos contained embedded quizzes and/or coding exercises. Lessons were released at 12am each day, from Monday to Thursday. Students were encouraged to create a schedule each day to watch these lessons.

There were ten face-to-face lectures, one at 11am and another at 12pm, Monday to Friday. Students were allowed to choose to attend one of these lessons per week. 5 lecturers were assigned to these face-to-face lectures, each lecturing two sessions on a given day. The format of the face-to-face lectures was a recap of the material covered in the online videos during the past week, followed by a discussion and question and answer session related to the section being covered, and then students were asked to complete a few coding exercises, and finally a discussion of the solution to the coding exercises. All face-to-face lectures were recorded and made available to all students.

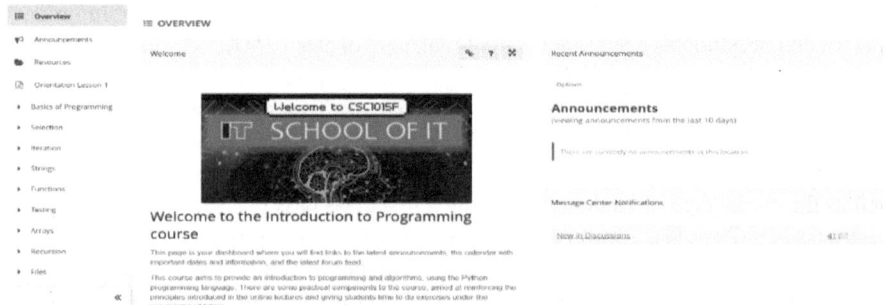

Fig. 5. Course homepage

Ten 2-h weekly practical sessions were also scheduled, starting at 2pm and 4pm each day. Due to lab sizes and availability (especially since lab capacities were restricted due to COVID restrictions), one classroom venue was booked for each session to accommodate the overflow of students. Students would bring their laptops to these venues. Each practical session had a number of student tutors assigned to the session, according to a ratio of approximately 1 tutor per 17 students. During the first 30 min of the practical session, students were required to complete a weekly quiz, which covered aspects of the online videos from the previous week.

Students were also allowed to interact with their tutors outside of the practical sessions via an online Discussion Forum, hosted on the LMS, as well as on WhatsApp chat groups. The Discussion Forums were divided into 4 Topics, i.e., Course Admin, Lectures (which were subdivided into each Section), Assignments (which were subdivided by Assignment, then by Question) and Quizzes. 5 WhatsApp chat groups were created, one for each practical day. Students were provided with the links to join these groups. Each group had about 12 tutors assigned to the group. Tutors were scheduled according to days when they would be responsible to answer any query that is posted on the chat. Strict rules were put in place so no "junk" messages were allowed, and no code could be posted on the chat. When students had queries on the assignment solutions, students would submit the code to the Automarker system and tutors would then access the code and provide feedback to the students.

The three practical tests were conducted during the weekly practical sessions, with questions uploaded on the LMS system and students were given a password to access the question during the practical session. Each test was assigned to specific groups (according to practical session sign-up), so questions were not visible to students outside of the practical session. Student solutions were submitted and automatically graded by the Automarker system.

4 Results and Discussion

4.1 Effect of Blended Learning on Final Marks

Students' overall performance was determined by their grades in the course, which is measured out of 100 and consists of the following: Weekly Assignments and Quizzes (15%), Theory Tests (15%), Practical tests (10%) and Examination (60%). In order to pass the course, students need to achieve a Final mark of at least 50%, as well as a Practical average (weighted average of Weekly Assignments, Quizzes and Practical Tests) above 45% and a Theory average (weighted average of Theory Tests and Examination) of above 45%. The students' overall pass rates for 2019 (the last time the course was offered in the traditional format, due to the COVID 19 pandemic) and for 2022 are shown in Table 2, as well as the percentage of students achieving a first class pass (1, >75), an upper second class pass (2+, 70 – 74), a lower second class pass (2−, 60 – 69), and a third class pass (3, 50 – 59). The pass rates for 2022 are slightly higher than 2019. While the number of first and upper second class passes seems to be lower in 2022, there is an increase in the number of lower second and third class passes.

Table 2. Student pass rates

	2019	2022
Pass	85,3%	88,0%
1	30,2%	26,1%
2+	13,2%	11,5%
2−	21,7%	24,8%
3	16,9%	25,6%

4.2 Feedback on Blended Learning

Further qualitative and quantitative analyses were carried out to examine the effectiveness of the Tashiv tool, the face-to-face lectures, some other components of the course and blended learning. 67 students responded to a post-questionnaire. Table 3 shows that the majority of students watched most or all of the videos. About half of the students indicated that they would watch the lesson videos before the face-to-face lectures (Table 4). Table 5 indicated the level of engagement of students with the coding exercises and inline quiz components of the Tashiv tool. From Table 5 we can see that students are engaging differently and also engaging with the components of the Tashiv tool differently. From this data, we see that students are choosing to engage with different tools, because they are now directing their own learning. By moving the learning from a traditional classroom into an online space, even with the use of equivalent tools, students are making choices of which tool to use, and when to use it and this is the advantage of the blended learning format.

Table 3. Number of videos watched

Number of videos watched	Responses
I watched all the videos	49 (73%)
I watched most of the videos	14 (21%)
I watched about half of the videos	0 (0%)
I watched very few of the videos	4 (6%)
Not at all	0 (0%)

Table 4. When was the lesson watched?

When was the videos watched	Responses
Before the face-to-face lecture	35 (52.2%)
After the face-to-face lecture	22 (32.8%)
Only before a test or exam	7 (10.5%)
Never	3 (4.5%)

Table 5. Level of engagement with components of the Tashiv tool

	Coding exercises	Inline quizzes
All of them	8	27
Most of them	11	11
About half of them	13	15
Very few of them	24	9
None of them	11	5

Table 6 shows the students opinions on how valuable the videos, the coding exercises and the inline quizzes were in learning the course material. We find that while there is a tendency towards students agreeing and strongly agreeing that the videos were valuable, we still find that some of the students don't agree that the videos were valuable. It may be that these are students who learn differently, or these are students who have prior knowledge of introductory programming. We find that there is more neutrality with regards to the value of the inline quizzes and the coding exercises. This suggests that students find more value in watching the videos than doing the exercises. This is an indication that in the Tashiv tool, students have agency to decide what to do. The fact that the students did not find these components important may have implications for how we use them in the Tashiv tool in future.

Table 6. Students' opinion on how valuable the Tashiv Tool components were in learning the course material

Was the following valuable?	Lecture videos	Quiz	Coding Exercises
1 (Strongly disagree)	6	7	3
2	5	8	12
3 (Neutral)	6	25	25
4	13	15	16
5 (Strongly agree)	37	12	11

51% of students indicate that they watched the videos again when preparing for the tests and examinations, while less than 25% of students did the quizzes and coding exercises again. Again, we find that there is variability in the use of the tools, but we also see that there is a stronger connection between the students and the videos as compared to the other components of the Tashiv tool.

In free form responses from students, two themes emerged: 1. There were some issues with the Tashiv tool as this was the first time it was rolled out, such as "Some of the coding aspects didn't work." and "Some exercises didn't work properly. But it mostly didn't hinder much the learning experience."; and 2. Some of the students are asking for more structure within the tool, such as "The user must not be able to answer on Tashiv tool after the person has finished the video." One student did comment that "I found lecture videos were infinitely more useful in learning than the in-person lectures." More than 81% of the students felt that the Tashiv tool was easy to use.

Table 7 shows the students' opinions on the value of the face-to-face lectures as a whole, as well as the subsections of the lectures. From Table 7, we also find a strong preference for passive learning. Students tend to prefer the recap of the online lessons and not want to engage in class activities. When asked to comment on the face-to-face lectures, 18 students responded and three trends emerged: some students wanted traditional lectures, some felt no value in the face-to-face lectures and some students wanted more active engagement. This shows that students have different learning styles.

Table 8 shows the students' opinions on how valuable the Discussion Forums, weekly quizzes, weekly assignments, and practical tests were in learning the course material.

Students' opinion on the discussion forums was largely neutral, while the tendency for the WhatsApp tutor groups was because it was more valuable to their learning. Most students agreed that the weekly assignments and the practical tests were valuable for their learning, while the opinion was not as strong with regard to the weekly quizzes. This indicates that even though the format of the course is changing quite substantially, there is still a strong emphasis on the assignments and the students understand the importance of the assignments.

Table 7. Students' opinion on how valuable the face-face lessons were in learning the course material

Was the following valuable?	Face-to-face lessons	Recap	Question and Answer	Coding exercises
1 (Strongly disagree)	4	0	3	3
2	17	5	10	7
3 (neutral)	13	8	17	14
4	11	22	18	18
5 (Strongly agree)	17	27	14	20

Table 8. Students' opinion on how valuable the other aspects of the course were in learning the course material

Was the following valuable?	Discussion forums	WhatsApp tutor groups	Weekly assignments	Weekly quizzes	Practical Tests
1 (Strongly disagree)	9	0	1	3	0
2	8	4	1	1	2
3(neutral)	29	18	1	10	7
4	9	12	9	15	10
5 (Strongly agree)	2	23	45	28	38

Students were asked to rank the aspects of the course in terms of importance. The summary of the ranking is shown in Table 9. This confirms that students felt the videos were the most important, followed by the assignments. The weekly quizzes and WhatsApp tutor groups also ranked high on the list. On the other end of the ranking, we find that the discussion forums and the hotseat were ranked as the least important part of the course. This is interesting as, in many courses, the discussion forum is put forward as the primary learning tool, but we notice that when students have options, they prefer other means of learning than the normative tool that is made available.

Table 9. Ranking of aspects of the course

	1	2	3	4	5	6	7	8	9
Tashiv tool: Lecture videos	38	3	3	3	4	1	2	0	1
Tashiv tool: Quizzes	0	3	1	5	9	11	9	14	3
Tashiv tool: Coding Exercises	0	3	11	2	6	10	13	6	3
Face-to-face lectures	4	7	4	8	12	7	5	1	6
WhatsApp tutor groups	1	3	4	18	6	6	5	6	4
Vula Discussion Forums	0	0	2	1	4	6	6	12	20
The Hotseat	1	5	1	4	7	5	8	10	11
Weekly Quizzes	2	7	13	14	6	5	3	3	3
Weekly Assignments	10	25	17	0	0	3	1	0	0

Of the students who responded, Fig. 6 shows the students' opinion on the level of difficulty of the course, with 90% of students indicating a level of satisfaction with the course.

Fig. 6. Level of difficulty in this course

Students indicated that the aspects of the course that they liked most were the lecture videos and the weekly assignments, while some students indicated they experienced some level of difficulty in completing the weekly assignments. In relation to other courses (which still adopted the traditional approach to teaching), 78% of students responded that the workload for this course was on average or slightly above average as other courses. Only 15% of students responded that the workload was a lot more than compared to other courses. In response to open ended questions, 85% of students who responded believed the blended learning format was better than the traditional approach to teaching the introductory programming course. Students liked the fact that lectures were online, and they could watch them at their own pace and watch them over and over again. Another important point made was that as students were transitioning to university life, students without prior knowledge of programming felt less intimidated by those students who had prior programming knowledge.

5   Conclusion

The demand for models of blended learning courses (which combine face-to-face lessons and online lessons) are becoming greater in all areas of academic education. The model of a blended learning course for teaching introductory programming at the University of Cape Town described here, developed out of the traditional model. The structure of this course has been enhanced to improve the conditions for learning in the given area and make them suitable for today's students.

The preparation and organization of the blended learning course required a significant amount of time since it was necessary to prepare and organize each of the numerous course segments (including the learning mechanism, new roles of the lecturer and student, course structure, new forms of knowledge base and learning environment, scenario, and the evaluation mechanism).

This paper studied the impact of using a blended learning method on an introductory programming course for the first time. The local online LMS system was used as a practical solution for the blended learning course. 1044 students enrolled in the introductory programming course. Their performance was comparable to traditional learning. Student satisfaction was also measured. The results revealed that students found the online system easy to use. We observe that the way in which students are engaging with material seems to be more passive video watching, rather than active engagement. This may be a result of the fact that these are the first cohort of students coming into University after the COVID 19 pandemic. It may be that these students were not getting the levels of engagement in schools for almost two years prior to coming to university. We also find that putting in a large number of tools that have been designed to operate as students expect gives students the ability to choose what they would use, as opposed to tools that the university prefers they use. The blended learning model also successfully reduced the need for staff by 60%.

In future work, we would like to study how we should design the online tools to foreground the active engagement or if the passive consumption of videos is sufficient. Also, blended learning will be applied to an advanced programming course taught at the university, such as Object-Oriented Programming using Java.

Acknowledgements. This research was funded by the University of Cape Town.

References

1. Cakiroglu, U.: Using a hybrid approach to facilitate learning introductory programming. Turk. Online J. Educ. Technol.-TOJET **12**(1), 161–177 (2013)
2. Dawson, J.Q., Campbell, A., Valair, A.: Designing an introductory programming course to improve non-majors' experiences (2018)
3. Alturki, R.A.: Measuring and improving student performance in an introductory programming course. Inform. Educ. **15**(2), 183–204 (2016)
4. Hassanzadeh, A., Kanaani, F., Elahi, S.: A model for measuring e-learning systems success in universities. Expert Syst. Appl. **39**(12), 10959–10966 (2012)

5. Wang, H.-Y., Wang, Y.-S., Shee, D.: Measuring e-learning systems success in an organizational context: scale development and validation. Comput. Hum. Behav. **23**(4), 1792–1808 (2007)
6. Holsapple, C.W., Lee-Post, A.: Defining, assessing, and promoting e-learning success: an information systems perspective. Decis. Sci. J. Innov. Educ. **4**(1), 67–85 (2006)
7. Garrison, D., Kanuka, H.: Blended learning: uncovering its transformative potential in higher education. Internet Higher Edu. **7**(2), 95–105 (2004)
8. Garrison, D., Vaughan, N.: Blended Learning in Higher Education: Framework, Principles and Guidelines. Wiley, Hoboken, NJ, USA (2008)
9. Singh, H.: Building effective blended learning programs. Educ. Technol.-Saddle **43**(6), 51–54 (2003)
10. Anggrawan, A., Ibrahim, N., Suyitno, M., Satria, C.: Influence of blended learning on learning result of algorithm and programming. In: 3rd International Conference Information Computing (ICIC), p. 1 – 6, October (2018)
11. Bati, T.B., Gelderblom, H., Van Biljon, J.: A blended learning approach for teaching computer programming: design for large classes in Sub-Saharan Africa. Comput. Sci. Educ. **24**(1), 71–99 (2014)
12. Jacobs, C.T., Gorman, G.J., Rees, H.E., Caig, L.E.: Experiences with efficient methodologies for teaching computer programming to geoscientists. J. Geosci. Educ. **64**(3), 183–198 (2016)
13. Ben-Ari, M.: Constructivism in computer science education. J. Comput. Math. Sci. Teach. **20**(1), 45–73 (2001)
14. Jenkins, T.: On the difficulty of learning to program. In: 3rd Annual Conference of the LTSN Centre for Information and Computer Sciences, Centre for Information and Computer Sciences (2002)
15. McCracken, M., et al.: A multi-national, multi-institutional study of assessment of programming skills of first-year CS students. ACM SIGCSEBull. **33**(4), 125–180 (2001)
16. Bennedsen, J., Caspersen, M.E.: Teaching object-oriented programming-towards teaching a systematic programming process. In: Proceedings of the 8th Workshop on Pedagogies and Tools for the Teaching and Learning of Object Oriented Concepts. Affiliated with 18th European Conference on Object-Oriented Programming (ECOOP), Oslo, Norway (2004)
17. Gomes, A., Mendes, A.: A teacher's view about introductory programming teaching and learning: difficulties, strategies and motivations. In: 2014 IEEE Frontiers in Education Conference (FIE) Proceedings. Madrid, Spain (2014)
18. Grissom, S.: Introduction to special issue on alternatives to lecture in the computer science classroom. ACM Trans. Comput. Educ. (TOCE) **13**(3), 9 (2013)
19. Dziuban, C., Graham, C.R., Moskal, P.D., Norberg, A., Sicilia, N.: Blended learning: the new normal and emerging. Int. J. Educ. Technol. High. Educ. **15**(1), 3 (2018)
20. López-Pérez, M.V., Pérez-López, M.C., Rodríguez-Ariza, L.: Blended learning in higher education: students' perceptions and their relation to outcomes. Comput. Educ. **56**(3), 818–826 (2011)
21. Alammary, A., Carbone, A., Sheard, J.: Curriculum transformation using a blended learning design toolkit. In: 40th HERDSA Annual International Conference, Sydney, Australia (2017)
22. Griffin, D.K., Mitchell, D., Thompson, S.J.: Podcasting by synchronising PowerPoint and voice: What are the pedagogical benefits? Comput. Educ. **53**(2), 532–539 (2009)
23. Vo, H.M., Zhu, C., Diep, N.A.: The effect of blended learning on student performance at course-level in higher education: a meta-analysis. Stud. Educ. Eval. **53**, 17–28 (2017)
24. Hadjerrouit, S.: Towards a blended learning model for teaching and learning computer programming: a case study. Inf. Edu. **7**(2), 181 (2008)
25. Yigit, T., Koyun, A., Yuksel, A.S., Cankaya, I.A.: Evaluation of blended learning approach in computer engineering education. Procedia-Social Behav. Sci. **141**, 807–812 (2014)

26. Grover, S., Pea, R., Cooper, S.: Designing for deeper learning in a blended computer science course for middle school students. Comput. Sci. Educ. **25**(2), 199–237 (2015)

27. Demaidi, M.N., Qamhieh, M., Afeefi, A.: Applying blended learning in programming courses. IEEE Access **7**, 156824–156833 (2019)

28. Chan, A.T.S., Cao, J., Liu, C.K., Cao, W.: Design and implementation of VPL: a virtual programming laboratory for online distance learning. In: International Conference on Web-Based Learn, Berlin, Germany (2003)

29. Oliver, K., Stallings, D.: Preparing teachers for emerging blended learning environments. J. Technol. Teach. Educ. **22**(1), 57–81 (2014)

30. Levine, S.J.: Making distance education work: understanding learning and learners at a distance. New Horiz. Adult Educ. Hum. Res. Dev. **25**, 124–126 (2013)

31. Laakso, M.: Promoting programming learning. Engagement, Automatic Assessment with Immediate Feedback in Visualizations. TUCS Dissertations no. 131 (2010)

32. Epstein, M.L., Epstein, B.B., Brosvic, G.M.: Immediate feedback during academic testing. Psychol. Rep. **88**(3), 889–894 (2001)

33. Stein, J., Graham, C.R.: Essentials for Blended Learning: A Standards-Based Guide. Routledge, Evanston, IL, USA (2014)

34. Kirkwood, A., Price, L.: Technology-enhanced learning and teaching in higher education: What is 'enhanced' and how do we know? A critical. Learn. Media Technol. **39**(1), 6–36 (2014)

AI and Future Movements;
Programming

Demystifying the Impact of ChatGPT on Teaching and Learning

Tapiwa Gundu[1]([✉]) [iD] and Colin Chibaya[2] [iD]

[1] Nelson Mandela University, Port Elizabeth, South Africa
tapiwag@mandela.ac.za
[2] Sol Plaatje University, Kimberley, South Africa
colin.chibaya@spu.ac.za

Abstract. The use of chatGPT as a teaching and learning tool is, generally, seen as a binary adventure to either destroy the traditional approaches to education or to innovatively revolutionize teaching and learning. Using large language models, chatGPT can generate detailed responses to prompts and follow-up questions. In this article, the hope is to explore the benefits and risks associated with using chat-GPT in teaching and learning, the opportunities it offers to students and instructors, as well as the challenges it brings about. The key paradoxes thereof are discussed before precise recommendations are separately listed for students and instructors. However, generally, chatGPT is a tool that can be used along with specific strategies for educational benefits.

Keywords: ChatGPT · Generative Artificial Intelligence

1 Introduction

Generative artificial intelligence (AI) refers to a type of artificial intelligence system that can generate new and original content based on the patterns it has learned from a given dataset [1]. The history of generative AI can be traced back to the early days of AI research. In recent years, the development of generative AI has accelerated due to the availability of large datasets and powerful computing resources. One of the most significant breakthroughs in generative AI has been the development of generative adversarial networks (GANs), which were introduced in 2014 by Ian Goodfellow [2]. GANs consist of two neural networks, a generator, and a discriminator, that are trained together in a process of competition and cooperation to generate high-quality, realistic outcomes [1]. Further developments gave birth to what is called transformer architecture [1]. This is a deep learning model architecture introduced in the paper "Attention Is All You Need" by Vaswani et al. in 2017 and has had a significant impact on natural language processing (NLP) tasks. It has become the foundation for various state-of-the-art models, including GPT (Generative Pre-trained Transformer). The transformer architecture is based on the concept of self-attention mechanisms, which allows the model to weigh the importance of different positions or tokens in the input sequence when processing each token [1]. This mechanism enables the model to capture dependencies between words

H. E. Van Rensburg et al. (Eds.): SACLA 2023, CCIS 1862, pp. 93–104, 2024.
https://doi.org/10.1007/978-3-031-48536-7_7

in a more flexible and efficient way compared to traditional recurrent neural networks or convolutional neural networks.

The main components of the transformer architecture are encoder, decoder, attention, and positional encoding. The encoder processes the input sequence and consists of a stack of identical layers. Each layer has two sub-layers: a multi-head self-attention mechanism and a feed-forward neural network. The self-attention mechanism captures the relationships between different words in the input sequence, while the feed-forward network applies a non-linear transformation to each position separately. The decoder generates the output sequence based on the encoded representation of the input. Similar to the encoder, the decoder is also composed of a stack of identical layers. In addition to the self-attention and feed-forward layers, the decoder also includes an additional attention mechanism called encoder-decoder attention. This attention mechanism allows the decoder to focus on relevant parts of the input sequence while generating the output. The attention mechanism is a key component of the Transformer architecture. It allows the model to weigh the importance of different positions in the input sequence when processing a particular position. The attention scores are computed based on a similarity function between the query (current position), key (all positions), and value (all positions). The self-attention mechanism in the encoder captures dependencies between words within the same input sequence, while the encoder-decoder attention in the decoder enables the model to consider the input sequence when generating the output. Since the transformer architecture does not have an inherent notion of word order like RNNs or CNNs, positional encoding is used to provide information about the order of words in the input sequence. Positional encodings are added to the input embeddings and provide the model with positional information.

The Transformer architecture has several advantages, such as parallelizability, which allows for more efficient training on modern hardware, and the ability to capture long-range dependencies in the input sequence. These properties have made the transformer architecture a popular choice for various NLP tasks, including machine translation, text summarization, question answering, and language generation.

ChatGPT is based on the transformer architecture [3] that has the potential to transform teaching and learning. It is a computational linguistic model designed to process and analyze natural language data using computational techniques. ChatGPT is a language translation, language generation, and text classification tool. However, its use in teaching and learning requires careful consideration of the risks and benefits it brings. This paper aims to demystify the impact of ChatGPT on teaching and learning by exploring its potential to either bring about a catastrophic downfall or facilitate transformative reformation, while offering a balanced perspective from the standpoint of instructors. In the conclusion section, the paper provides some recommendations for effective use of ChatGPT in education.

1.1 Statement of the Problem

The problem addressed in this study is an assessment of the merits and drawbacks of using chatGPT in education. Focus is invested on understanding the uses of chatGPT, the opportunities it offers to both the students and instructors, the challenges it presents to the same groups, as well as the paradoxes thereof. In the end, we share some recommendations for effective use of the tool to enhance teaching and learning.

1.2 Overview

The article proceeds as follows: Sect. 2 presents a thorough examination of related work using a desktop analysis of literature that considered the reliability of the information sourced from chatGPT. As the subject matter is relatively new, only a limited number of scholarly articles and non-peer-reviewed academic papers, such as white papers, were discovered as of April 2023. This limited literature relates to ChatGPT and its connection to higher education. In this review, emphasis was placed on understanding the chatGPT system, its application in teaching and learning, the opportunities it brings to students and instructors, as well as the downside of the system from both students and instructors' perspectives. This literature was searched from Google Scholar, soliciting prevalent academic articles, conference proceedings, and book chapters related to research on the use of chatGPT in higher education. We also checked the reference lists of certain academic articles and citations in non-academic articles. Through this comprehensive search strategy, we found out that there is limited academic literature on ChatGPT and higher education. In Sect. 3, we elucidate the paradoxes of chatGPT and the future of education. Section 4 shares our recommendations in view of the potential and challenges to both the instructors and students. We conclude the paper in Sect. 5, highlighting the key contributions, and the likely future works.

2 Related Work

The key research question in conducting this review interrogated literature on whether machines can think. Generally, it is insinuated that discussions on the topic of whether machines can or cannot think are still ongoing. The major debate is still around the definition of "thinking" and whether we believe that consciousness and subjective experience are essential components of this process. From a technical standpoint, machines can perform complex calculations and execute tasks based on predefined rules and algorithms, which can resemble human thinking [4]. However, this is fundamentally different from the way humans think and experience the world, which involves subjective consciousness, emotions, and creativity [5].

ChatGPT is a conversational AI interface that leverages natural language processing (NLP) to interact in a natural-sounding way [6]. It can answer follow-up questions, acknowledging mistakes, challenging incorrect assumptions, and declining inappropriate requests, among other features. Although ChatGPT was designed to emulate human conversation [7], its abilities extend far beyond that. It can generate entirely new works, such as poems, stories, or novels, as well as simulating virtually any behaviour within its capabilities.

When asked to explain itself, ChatGPT generated the following response: "ChatGPT is a large language model developed by OpenAI, based on the GPT (Generative Pre-trained Transformer) architecture. It uses deep learning techniques to generate human-like responses to text-based inputs. ChatGPT is trained on a vast corpus of text data from the internet, allowing it to understand and produce language in a wide range of styles and topics. It can be used for a variety of natural language processing tasks, including text generation, language translation, question answering, and more". Another prompt was utilized as a follow-up question, which asked about the mechanics or operations of the chatGPT. We deliberately phrased this follow-up question as "How does it work?" instead of "How does ChatGPT work?" in order to gauge its conversational abilities". The response was as follows: "ChatGPT works by using deep learning algorithms to analyze large amounts of text data and learn patterns in language usage. Specifically, it is based on the Transformer architecture, which uses attention mechanisms to focus on relevant parts of input text when generating responses". The model is understood to be continuously learning. It learns to predict the next word in a sequence based on the preceding words. Also, the model learns the relationships between words and the structure of language. The learned relationships between words are used to generate responses when chatGPT is given a question. Its attention mechanism is key in identifying relevant parts of the training data before it generates responses based on the identified patterns in the data. Given these descriptions, our study discusses the pros and cons of using chatGPT as a teaching and learning tool.

2.1 Use of chatGPT in Teaching and Learning

There are several uses of chatGPT in teaching and learning. Its capabilities to generate articles, stories, poems, and essays, as well as providing summaries, adjusting texts, and writing computer code, are very attractive features in teaching and learning. On querying chatGPT on its role in teaching and learning, its response was *".... Can be used in several ways, such as providing virtual personal tutoring, creating learning assessments, and enhancing pedagogical practices. For instance, ChatGPT can assist learners in answering questions, clarifying concepts, and providing feed-back. It can also help educators in designing course content, generating exam questions, and identifying knowledge gaps......"* [8]. However, some educationists and researchers have expressed concerns regarding ethics and academic integrity. Although some see it as a revolutionary tool for teaching and learning, representing the future of education, others view it as a threat to traditional educational practices, a danger to education. The bulk of analysts see it as hindering the development of critical thinking skills in students. They think that it challenges academic integrity, bringing about bias and over-reliance on technology. This article seeks to establish clear guidelines and best practices for ethical and effective use of chatGPT in teaching and learning.

2.1.1 Opportunities to Students

Several opportunities to students are apparent. Commonly, students lag in engaged conversational learning. ChatGPT can facilitate conversational learning by allowing students to interact with the AI language model in a natural and intuitive way. It provides an interactive communication platform that allows students to follow more engaging classroom activities. Students can produce visual aids such as slides or worksheets that clearly articulate lesson objectives and success criteria. These instructional materials have a high potential to attract students and encourage them to participate in classroom learning. Additionally, chatGPT-generated questions and prompts can stimulate students' problem-solving and critical thinking abilities at various levels of knowledge and skill [9] which is crucial for modern-day education.

Also, a generative model such as chatGPT can aid in the creation of adaptive learning systems that adapt their teaching techniques according to a student's advancement and achievements. Adaptive learning systems such as chatGPT are effective in assisting students in complex modules, e.g., the learning of programming [10], resulting in better assessments scores. The model's capability to comprehend students' expertise levels allows it to produce problems of suitable difficulty [10].

More so, chatGPT is a potential solution that can serve as a personal tutor for students, unlike traditional tutoring methods. Students can receive individual feedback and answers by interacting with ChatGPT [11], which provides intellectual tutoring services virtually [12]. This feature allows students to seek assistance from ChatGPT at any time, be it with their homework, assignments, projects, or even math exercises. This approach may empower students to become autonomous and self-directed learners. Its capability to allow students to engage in discussions on various topics, in addition to answering their questions strengthens students independence [1]. It enhances students' research capabilities [9]. In this case, students can provide prompts to ChatGPT and request it to generate outlines for dissertations or other forms of writing [9]. Thus, students can better organize their ideas for research and writing [1, 11, 12], allowing them to comprehend the main points quickly and accurately [12]. However, students would require individual abilities to adjust some of the content for the outline to be usable.

One common challenge in higher education is affording equal accessibility to study resources to a diverse population of students. ChatGPT can provide learning opportunities for students with disabilities or who are unable to attend in-person classes due to distance or other factors. Equally, educational materials can be translated into different languages, making them accessible to a wider audience [13]. The model can achieve outstanding results on several translation benchmarks, indicating its ability to comprehend words in a language and generate translations in another language [13].

In addition, that timeliness and promptness in giving students immediate feedback, allowing students to receive immediate responses to their queries or assignments is an attractive feature in education. Timely feedback enables students to understand and correct their mistakes quickly, which can lead to faster learning and improvement. Also, receiving quick feedback can help students stay motivated and engaged in their learning, as they can see the results of their efforts, encouraging them to continue learning. However, for students to benefit optimally from using chatGPT, institutions should establish clear guidelines and ensure that students properly, effectively, and responsibly use the tool.

2.1.2 Opportunities to Instructors

ChatGPT can also assist instructors with automated administrative tasks such as grading, feedback, and providing answers to frequently asked questions from students. This allows instructors to focus on more important tasks such as teaching and research [6]. The system can be programmed to evaluate student essays [11]. It can evaluate beginner and intermediate essays accurately by training it on a dataset of essays that were previously graded by humans [3], effectively recognizing critical features of well-written essays against badly written works.

Similarly, chatGPT can provide instructors with tools to create more effective learning experiences for their students, providing immediate feedback on student work, identifying areas of difficulty, and suggesting personalized resources to address their needs. This way, instructors can create content such as lectures, tutorials, and assignments. However, it is important to note that the instructor should review and edit the content generated to ensure accuracy and relevance. Since assessment is a crucial component of high-quality education, chatGPT offers a flexible and innovative method to create learning assessments that provide real-time feedback and reports [1]. Instructors often spend a significant amount of time creating quizzes, monthly tests, and exams. Also, chatGPT can generate multiple choice questions based on a set of criteria provided by the instructor. These features can save time on creating assessments, while creating learning assessment items that adhere to a standard framework, potentially improving the quality of the questions, as well as ensuring that the questions are relevant and aligned with the learning objectives [14]. This way, instructors can develop open-ended questions that align with the success criteria of the teaching lessons [1]. The capability of chatGPT to create adaptive assessments that adjust the difficulty level based on the student's performance is also very attractive. This can help to ensure that students are challenged at an appropriate level and that their learning needs are met. Hopefully, aiding in creating assessments, chat GPT enables instructors to decrease their workload and prioritize the development of innovative lesson plans, participation in professional development programmes, and providing coaching and mentoring support to students individually. These activities are crucial for enhancing students' learning performance.

2.1.3 Challenges to Students

While several opportunities for students are apparent, there are key drawbacks to note. Obviously, chatGPT cannot provide the same level of human interaction as instructors. This lack of human interaction can be a disadvantage for students who may benefit more from a personal connection with the instructor. Students who interact with real instructors, often, have a better learning experience [15].

Another drawback is that chatGPT relies on statistical patterns in their training data and lack true understanding of the concepts being taught. This limitation can hinder their ability to provide personalized explanations or feedback that address a student's individual needs and misconceptions [1]. In fact, tutoring systems based on chatGPT models may not provide customized explanation to students' misconceptions [1].

On the other hand, there is a risk that chatGPT may produce misleading information, which may negatively affect both the instructor and students. Several limitations can be pinpointed, including the potential for factual errors such as generating articles that do not exist or providing inaccurate responses [12]. False information may also arise due to inadequate input, insufficient training data, or from chatGPT's limitations. For example, chatGPT has a limited capacity to comprehend context, and produce responses that are appropriate to the situation of a conversation [16].

Students may also become overly dependent on AI, which could pose a problem. Students, especially those who procrastinate, may rely solely on the tool to create their work without engaging their critical thinking and decision-making skills. This can negatively impact the development of critical thinking skills [9, 11, 14, 17], problem-solving skills [9], as well as imagination and research abilities [17]. Given that these abilities are crucial to academic and professional success, depending solely on ChatGPT could have several adverse effects on students, including a lack of creativity and weak decision-making skills.

To avoid misunderstandings, it is crucial for instructors and students to critically evaluate the information generated by ChatGPT. It is important to use AI-generated information effectively and efficiently, exercising caution where necessary.

2.1.4 Challenges to Instructors

The shortcomings of using chatGPT also pose significant risks to instructors. Its ability to generate outputs based on various prompts submitted by users, particularly students, may be used unethically and against the standard of academic integrity [18]. Unless its use is declared or the sources are appropriately cited, its utilization may result in plagiarism [18, 19]. If students do not acknowledge sources, it may lead to academic misconduct and dishonesty, which ultimately harms their academic profile.

Use of chatGPT also raises concerns about potential bias in evaluating students' work. It can be difficult to distinguish between the output generated by chatGPT and that produced by humans [17, 20, 21]. That rapid and accurate essay generation capability of chatGPT has led to concerns among instructors that students may use it to outsource their assignments and projects, potentially resulting in higher scores compared to their peers [11]. One Twitter user reported that students used chatGPT to write school essays and received A+ grades. Another twitter established that chatGPT essays scored over 70% on a practice bar exam [22]. Such incidents could have a negative impact on students' emotions and the reputation of educational institutions. Therefore, awareness and knowledge of potential bias is key.

3   The Paradoxes of chatGPT and the Future of Education

Discussion regarding the use of chatGPT in education are, often, more centered on its challenges [6, 16, 17, 23] or the opportunities it offers [9, 12, 14, 21]. These opposing views depict the paradoxical nature of chatGPT and its role in education. To be precise, chatGPT has the potential to undermine and enhance certain educational practices simultaneously. To further explore these conflicting ideologies, we discuss four key paradoxes of ChatGPT in education, presenting hands-on practical examples, which, in turn, offer useful lessons and implications for the future of higher education.

- Authenticity vs. Quick Response: a balance is needed between quick response and the authenticity of results. ChatGPT can provide students with quick feedback on their work, which can be beneficial for their learning. Students can get immediate feedback on their progress and performance. This can help students identify areas that require further scrutiny, areas where they need to improve and adjust their learning strategies accordingly. However, quick responses may not always be authentic or accurate. Instructors and students should strike a balance between these two likely outcomes in using chatGPT.

- Exclusion vs Access: chatGPT enhances access to education. However, it also has the potential to exclude those who lack access to the technology required to run it. Access and exclusion are two critical issues in the education field that often exist in tension with one another. Access refers to the ability of individuals to participate in and benefit from education, irrespective of their background, socioeconomic status, or location. It is a way to promote social justice and equal opportunities. Exclusion, on the other hand, refers to the ways in which certain individuals or groups may be marginalized within the education system. Exclusion can take various forms, such as lack of resources or support, discrimination, or bias, or cultural or linguistic barriers. This potential to alleviate educational disparities or disadvantaged certain groups is noteworthy. Instructors can achieve a tradeoff by generating materials in various languages, making them accessible to students who speak languages other than the language of instruction, and customizing educational resources to meet the specific linguistic and cultural needs of diverse communities. Nonetheless, there are also concerns about the possibility of chatGPT to perpetuate existing biases and inequalities within the education system. For instance, if it is trained using biased or limited datasets, it may generate content that reinforces existing prejudices and stereotypes. Additionally, there is a risk that it could replace human educators, resulting in job loss and further amplifying inequities in education. Awareness about the functionalities of this tool to all stakeholders can demystify its impact.

- Objectivity vs bias: to promote equity and fairness in education, the use of chatGPT can provide objective assessments. However, it can also introduce biases if it is trained on limited or biased data. The balance between bias and objectivity is a crucial factor in education. Bias can hinder the promotion of inclusivity, diversity, and equity, by allowing personal or cultural perspectives, beliefs, and values to influence education. On the other hand, objectivity allows for a neutral and unbiased approach to education, free from any cultural or personal influences. Objectivity is essential to ensuring fair and equitable education that encourages critical thinking and evidence-based reasoning. ChatGPT has the potential to mitigate the tension between bias and objectivity by introducing a more objective and data-driven approach to education. It can analyze large datasets of student performance and learning outcomes to identify patterns and insights that inform educational practices and policies. Moreover, it can develop unbiased assessments and evaluations based on objective criteria, rather than subjective judgment. Nevertheless, the risk arises when the AI is trained on biased or limited datasets, generating content or assessments that reinforce existing stereotypes and biases. Additionally, the automation of certain educational aspects may restrict the educators' ability to detect and address bias in their teaching practices. Therefore, it is essential to ensure that the design and implementation of chatGPT prioritizes ethical and equitable practices and consider the potential for bias and other unintended consequences.

- Tradition vs innovation: use of chatGPT in education presents a tension between innovation and tradition. ChatGPT represents a novel approach to teaching and learning that challenges traditional methods and may even supplant the role of human instructors. Innovation and tradition are two core values in education that are often at odds with each other. Innovation involves the adoption of emerging technologies, practices, and approaches to enhance the quality, accessibility, and relevance of education in response to a rapidly changing world. On the other hand, tradition entails preserving established practices, beliefs, and values in education to maintain stability, continuity, and cultural heritage. ChatGPT has the potential to foster innovation in education, offering novel ways to develop and deliver educational content and assessments, and facilitating personalized and adaptive learning experiences. For instance, it could be leveraged to design interactive and immersive educational simulations, to provide tailored feedback and recommendations to students, and to generate new forms of educational content that are customized to meet the specific needs and interests of each learner. On the contrary, it may undermine or disrupt traditional educational practices and values. If deployed to automate certain aspects of education, such as grading or assessment, it may diminish the role of educators in the teaching and learning process and reduce opportunities for human interaction and engagement in education. Moreover, if it is used to provide personalized learning experiences, it may prioritize individualized learning goals and preferences at the expense of broader educational objectives and societal needs. Thus, it is crucial to strike a balance between its adoption in education and consideration of the potential impact on traditional educational practices and values. It is essential to use chatGPT in a manner that aligns with the broader goals of education, such as promoting critical thinking, fostering creativity, and nurturing social responsibility.

4   Recommendations

4.1   Recommendations for Students

Five recommendations are sequentially listed on the safe use of chatGPT by students:

1. Remember that chatGPT is a tool and not a substitute for human interaction. While it can provide quick and convenient answers, it is still essential to seek advice from your instructors, peers, or academic advisors when needed.
2. Be mindful of the information you provide to chatGPT, especially when discussing personal or sensitive matters. ChatGPT may store this information. It is essential to be cautious with the data you share.
3. Use chatGPT responsibly and avoid solely relying on it. Ensure that you are still developing critical thinking skills by double-checking the accuracy of its responses and seeking out additional sources of information.
4. Be aware of potential biases in chatGPT's responses, as it may have been trained on biased or incomplete datasets. If you suspect that chatGPT's response may be inaccurate, it's best to seek other opinions.
5. Remember that ChatGPT is not infallible and may not be able to answer all your questions or provide demonstrations of practical skills. Be sure to use it as a supplement to your studies and not a replacement for them.

4.2   Recommendations for Instructors

Five recommendations are tabled for instructors as follows:

1. ChatGPT can be a useful tool to enhance your lectures and provide quick answers to your students. However, it should not be relied upon. Use it solely as a supplement to your teaching, not as a replacement.
2. Be aware of the potential biases in the responses you receive and always verify the answers because chatGPT may have been trained on biased or incomplete datasets. If you notice any inaccuracies, it's best to seek out additional sources of information.
3. Encourage students to use chatGPT responsibly and develop their critical thinking skills. Emphasize the importance of double-checking the accuracy of its responses and seeking additional sources of information.
4. Educate students on the potential privacy and security risks associated with using chatGPT, such as the collection and storage of personal data. It's crucial to ensure that students are aware of the risks and take steps to protect their information.
5. Implement strategies to prevent cheating, such as timed assessments or proctoring measures. While chatGPT can be a helpful tool for students, it should not be used to cheat on assessments or exams. Integrating chatGPT with learning management systems which can include additional measures to deter cheating, such as plagiarism detection software should be encouraged.

5   Conclusion

Acknowledging the listed paradoxes is crucial. It is essential to thoroughly assess the advantages and drawbacks of utilizing chatGPT in education. Adopting a balanced approach that capitalizes on the benefits of chatGPT while acknowledging its shortcomings can help ensure its usage to promote equitable and effective education. We cannot ignore the advantages of using chatGPT over human instructors in terms of availability and accessibility. Unlike human instructors, chatGPT can be accessed by students at any time and from anywhere, providing round-the-clock support for their learning needs. It can provide immediate feedback on learning progress, allowing students to identify areas that require more focus and improve their understanding of the work. We cannot overlook the interactive learning experiences that students can gain through chatbots and these virtual assistants that can make learning more engaging, enhancing retention of information. Additionally, we cannot sideline chatGPT's benefits related to saving time for students and instructors by providing quick answers to questions, which frees up time for other academic pursuits and extracurricular activities.

On the other hand, despite its many benefits, we should remain aware that chatGPT lacks emotional intelligence and personal touch that human instructors offer. It lacks emotional support and guidance for students facing academic or personal challenges. We should be reminded that chatGPT's responses may be biased. Students should not rely too heavily on it because they may overlook the importance of human interaction and critical thinking skills. Also, chatGPT is not equipped to handle all types of queries and may not be able to provide practical demonstrations of skills.

Instructors cannot ignore an additional resource for swiftly generating study materials for their students. ChatGPT can also analyze vast amounts of data, providing valuable insights into students' learning progress and identifying areas where additional support may be needed. We cannot undermine the capabilities of chatGPT as a professional development tool for enhancing knowledge and understanding of different subjects areas. To be precise, this is not a tool for binary choices between destruction or improvement, but rather an equilibrium between the merits and drawbacks.

The following contributions emanate from this study:

(a) The synthesis of the uses of chatGPT in education, the opportunities to students and instructors, the drawbacks attached to its use in education, the paradoxes thereof, and the recommendations provided a comprehensive understanding of the merits and demerits of pursuing the use of this tool in teaching and learning. The discussions we shared in this study can help policymakers make informed decisions about undertaking such initiatives and interventions.

(b) The recommendations shared in this study, both to students and instructors, can help in avoiding common pitfalls and ensure that stakeholders maximize the potential benefits offered by the tool. The same recommendations also sought to ensure that proper use is emphasized in a way that maximizes its potential benefits for students and instructors to increase throughput.

Two directions for future research are planned as follows:

(a) We could investigate the long-term impact of the use of chatGPT to student outcomes, such as employability and career advancement. This investigation can help institutions assess the effectiveness of generative AI tools in education and make necessary adjustments.

(b) We could also explore the challenges and barriers that institutions face when implementing the initiatives to use chatGPT in their programmes. This can help institutions develop strategies to overcome these challenges and ensure that their initiatives are successful.

References

1. Baidoo-Anu, D., Owusu Ansah, L.: Education in the era of generative artificial intelligence (AI): understanding the potential benefits of ChatGPT in promoting teaching and learning, SSRN Electron. J. (2023). https://papers.ssrn.com/abstract=4337484. https://doi.org/10.2139/ssrn.4337484

2. Goodfellow, I.: NIPS 2016 Tutorial: Generative Adversarial Networks. SSRN Electron. J. (2017). https://doi.org/10.48550/arXiv.1701.00160

3. Choi, J.H., Hickman, K.E., Monahan, A., Schwarcz, D.: ChatGPT Goes to Law School. SSRN Electron. J. (2023). https://papers.ssrn.com/abstract=4335905. https://doi.org/10.2139/ssrn.4335905

4. Kowalski, R.: Computational Logic and Human Thinking: How to Be Artificially Intelligent. Cambridge University Press (2011). https://doi.org/10.1017/CBO9780511984747

5. Minsky, M.: The Emotion Machine: Commonsense Thinking, Artificial Intelligence, and the Future of the Human Mind. Simon and Schuster (2007)

6. Sok, S., Heng, K.: ChatGPT for education and research: a review of benefits and risks. SSRN Electron. J. (2023). https://papers.ssrn.com/abstract=4378735. https://doi.org/10.2139/ssrn.4378735

7. AI is now everywhere | British Dental Journal. https://www.nature.com/articles/s41415-023-5461-1. Last accessed 17 Apr 2023
8. ChatGPT: Generative artificial intelligence (AI) (2022)
9. Kasneci, E., et al.: ChatGPT for good? On opportunities and challenges of large language models for education. Learn. Individ. Differ. **103**, 102274 (2023). https://doi.org/10.1016/j.lindif.2023.102274
10. Alam, A.: Employing adaptive learning and intelligent tutoring robots for virtual classrooms and smart campuses: reforming education in the age of artificial intelligence. In: Shaw, R.N., Das, S., Piuri, V., Bianchini, M. (eds.) Advanced Computing and Intelligent Technologies: Proceedings of ICACIT 2022, pp. 395–406. Springer Nature Singapore, Singapore (2022). https://doi.org/10.1007/978-981-19-2980-9_32
11. Mhlanga, D.: Open AI in education, the responsible and ethical use of ChatGPT towards life-long learning. SSRN Electron. J. (2023). https://papers.ssrn.com/abstract=4354422. https://doi.org/10.2139/ssrn.4354422
12. Qadir, J.: Engineering Education in the Era of ChatGPT: Promise and Pitfalls of Generative AI for Education (2022)
13. Jiao, W., Wang, W., Huang, J., Wang, X., Tu, Z.: Is ChatGPT A Good Translator? Yes With GPT-4 As The Engine (2023). https://doi.org/10.48550/arXiv.2301.08745
14. Zhai, X.: ChatGPT user experience: implications for education. SSRN Electron. J. (2022). https://papers.ssrn.com/abstract=4312418. https://doi.org/10.2139/ssrn.4312418
15. D'Mello, S., Craig, S., Witherspoon, A., Graesser, A.: Affective and learning-related dynamics during interactions with an intelligent tutoring system. Int. J. Hum.-Comput. Stud. **72**, 415–435 (2014)
16. Duha, M.S.U.: ChatGPT in education: an opportunity or a challenge for the future? TechTrends (2023). https://doi.org/10.1007/s11528-023-00844-y
17. Shiri, A.: ChatGPT and academic integrity. SSRN Electron. J. (2023). https://papers.ssrn.com/abstract=4360052. https://doi.org/10.2139/ssrn.4360052
18. Kleebayoon, A., Wiwanitkit, V.: Artificial Intelligence, chatbots, plagiarism and basic honesty: comment. Cell. Mol. Bioeng. (2023). https://doi.org/10.1007/s12195-023-00759-x
19. Thurzo, A., Strunga, M., Urban, R., Surovková, J., Afrashtehfar, K.I.: Impact of artificial intelligence on dental education: a review and guide for curriculum update. Educ. Sci. **13**, 150 (2023). https://doi.org/10.3390/educsci13020150
20. Cotton, D.R.E., Cotton, P.A., Shipway, J.R.: Chatting and Cheating. Ensuring academic integrity in the era of ChatGPT, (2023). https://edarxiv.org/mrz8h/. https://doi.org/10.35542/osf.io/mrz8h
21. Else, H.: Abstracts written by ChatGPT fool scientists. Nature **613**, 423 (2023). https://doi.org/10.1038/d41586-023-00056-7
22. Taecharungroj, V.: "What can ChatGPT do?" analyzing early reactions to the innovative AI Chatbot on Twitter. Big Data Cogn. Comput. **7**, 35 (2023). https://doi.org/10.3390/bdcc7010035
23. de Winter, J.C.F.: Can ChatGPT pass high school exams on English Language Comprehension? Int. J. Artif. Intell. Educ. (2023). https://doi.org/10.1007/s40593-023-00372-z

Using ChatGPT to Encourage Critical AI Literacy Skills and for Assessment in Higher Education

Cheng-Wen Huang⬤, Max Coleman⬤, Daniela Gachago⬤,
and Jean-Paul Van Belle(✉)⬤

University of Cape Town, Rondebosch, South Africa
`jean-paul.vanbelle@uct.ac.za`

Abstract. Generative AI is about to radically transform the way intellectual and creative work is being done. Since the release of ChatGPT in late 2022, the potential impact of generative AI tools on higher education has been intensively debated. ChatGPT can generate well-formulated human-like text passages and conversations that is often, but not always, of a surprisingly high quality. This paper reports on an early experiment to explore ways in which ChatGPT can be used in the higher education context. The experiment involved a written assignment which required postgraduate Information Systems students to formulate a critique of the outputs of ChatGPT to a specific question in Information Systems project management. The paper investigates the level of criticality that the students demonstrated in working with ChatGPT and assessing the quality of its outputs. It further explores the claim that ChatGPT can be used to generate rubrics and assess students' assignments by asking ChatGPT to produce a rubric for critical thinking and assess the students' assignments against the rubric produced. The findings indicate that students perceive the ChatGPT produced responses as generally accurate, although they tend to lack depth, with some key information omitted, produced biased responses and have limitations with academic writing conventions. The rubric that ChatGPT produced for assessing critical thinking is lacking in certain areas and the reliability of using it as an assessment tool is questionable given the inconsistency in the results. Overall, the paper concludes that while ChatGPT and other text generative AI can be useful learning and teaching companions for both students and lectures, human expertise and judgement is needed in working with ChatGPT.

Keywords: ChatGPT · Teaching Tools · AI in Education · Generative AI · ICT in Education

1 Introduction

Since the release of ChatGPT in late 2022, the potential impact of generative artificial intelligence (AI) tools on higher education has been intensively debated. ChatGPT, which stands for Chat Generative Pre-training Transformer, is a software in the form of a chatbot that is capable of generating well-formulated human-like text passages and conversations that is often, but not always, of a surprisingly high quality. While some are dubbing

H. E. Van Rensburg et al. (Eds.): SACLA 2023, CCIS 1862, pp. 105–118, 2024.
https://doi.org/10.1007/978-3-031-48536-7_8

ChatGPT as the "greatest cheating tool ever invented" [12], others are optimistic about the possibilities the tool has for education, such as, providing personalized feedback and guidance to students, and helping educators reduce their workload through, for example, assisting with grading and lesson planning [7, 8, 10]. Within a relatively short period since the release of ChatGPT, an advanced, paid version, GPT-4, is already out in the market as of March 2023. Generative AI is a rapidly developing field and exploratory research is needed to understand the possibilities and pitfalls of its use in education.

This paper reports on an early experiment to explore ways in which ChatGPT can be used to foster students' critical thinking and assist lecturers with rubric making and grading in a case of a post-graduate Information Systems course. As such, the two research questions guiding this paper are:

1. What are students' perceptions on the usefulness of ChatGPT's outputs?
2. How does ChatGPT perform in producing a rubric and evaluating students' assignments?

The paper begins with a literature review that outlines the strengths and limitations of ChatGPT, as well as the opportunities and challenges the tool poses to higher education. It then introduces the research methodology and the student assignment on which the experiment is based. Corresponding with the two research questions above, the findings reveal that students perceive the ChatGPT produced responses as generally accurate, although they tend to lack depth, omit key information, exhibit biases and struggle with academic writing conventions. The rubric that ChatGPT produced for assessing critical thinking demonstrates some shortcomings and its reliability as an assessment tool is questionable due to inconsistent results. The paper concludes that while ChatGTP might be a useful thinking, learning and teaching companion, its meaningful integration in education requires conversation, engagement, critical thinking and analysis of ChatGPT's outputs.

2 Literature Review[1]

2.1 Strengths and Limitations of ChatGPT

Launched as a prototype by OpenAI in November 2022, ChatGPT is one of the most prominent text-generative AI tools belonging to the large language models (LLMs) category that is currently free to the public. Large language models are built on deep learning techniques that are trained on massive amounts of text data to learn the statistical patterns and structures of language. ChatGPT is based on the GPT-3 model which has 175 billion parameters (values that a neural network can optimize during training) [17]. GPT-3's ability to deal with complex information is evidenced in its passing of the U.S Bar exam, which typically takes students a decade to achieve [3].

ChatGPT is designed to process and generate human-like texts. It is capable of receiving prompts, generating outputs in real-time, engaging in human-like conversations and providing personalized responses building on past prompts [17]. Some language tasks it is capable of performing include summarizing, translating and explaining grammar

---

[1] Parts of this literature review are from the unpublished honours thesis by one of the authors [6].

issues [11]. Its outputs are noted to demonstrate a remarkable level of logical coherence and organization [9]. Thus, ChatGPT is seen to have the potential to enhance human efficiency through, for example, streamlining workflows, automating mundane tasks and improving operational efficiency.

LLMs are strongly dependent on the data they are trained [1]. ChatGPT's training data consists of publicly available sources, such as Common Crawl, Reddit Links and Wikipedia, up until the year 2021 [17]. For this reason, its outputs may not be the most up-to-date and may reflect the inaccuracies or biases that are present in the publicly available dataset [11]. Moreover, as a language model, ChatGPT produces its outputs based on next word predictions. This means that it has no real understanding, thus its outputs may not be fully attuned to the intricate nuances and subtleties of a given context [17]. Additionally, ChatGPT has been observed to have limited understanding and generation capabilities for low resource languages, such as, Nepali (a language native to the Himalayas), Sinhala (a language spoken in Sri Lanka), Khmer (the national language of Cambodia) and Pashto (a language native to Afghanistan and Pakistan) [2]. Its limitations in producing in-text and end-of-text references, and even going as far as to fabricate sources, have also been noted [17].

2.2   Opportunities that Generative AI Technologies & LLMs Present to Higher Education

ChatGPT's emergence has incited a flurry of deliberations and dialogues in the higher education space. A study of sentiments in the early stages found that early adopters of ChatGPT in education are generally positive, with many seeing its potential to bring about a paradigm shift in approaches to instruction and driving learning reform [19]. Certainly, ChatGPT's revolutionary impact in education has been likened to that of the calculator, which is now an indispensable tool in the field of Mathematics and everyday life in general [17].

Various recommendations have been put forward on how ChatGPT can be leveraged in higher education. These include, for example, using it as a quick reference or self-study tool, as indicated in a study with pharmacology students [13]. Kasneci et al. [11] propose that LLMs can be used as a writing assistance, helping to create summaries and outlines, which, in turn, can aid in grasping key concepts and organizing one's thoughts for writing. They also propose that LLMs can be used to cultivate research skills by not only offering information and resources, but also identifying current research as well as gaps. For teachers they suggest that LLMs can be used to create lesson plans and formative assessments, such as, practice problems and quizzes. ChatGPT's potential to save time with creating assessments is exemplified in case where a 100% productivity increase with exam-writing procedures was observed [1].

2.3   Challenges that Generative AI Technologies & LLMs Pose to Higher Education

Despite the opportunities, there are also concerns with the risks that ChatGPT poses to higher education. One key concern centers around the potential threat to academic integrity. ChatGPT has been shown to be able to pass various exams, from MBA [16], to law [3] to medication exams [5]. Thus, there are legitimate concerns that ChatGPT can lead to increases in cheating and plagiarism and ultimately undermine traditional assessment methods.

In particular, ChatGPT's emergence has raised significant doubts regarding the viability of the traditional essay as an assessment method. This concern has even led to the resurgence of the term 'the death of the essay'. For example, a paper which provides evidence of ChatGPT being able to achieve First-Class grades in a physics essay, is titled *The death of the short-form physics essay in the coming AI revolution* [20]. Although the results of this study demonstrate a threat to the integrity of short-form essays, they argue that this, in fact, presents an opportunity to *"instigate dramatic and profound changes to the way that we teach and assess students, forming an indispensable component of a new ethos within which we design and deliver teaching"* [20:10].

Other concerns that have been raised include the possibility of students and teachers becoming too reliant on ChatGPT that they come to lose their critical thinking and problem solving skills [11]; ethical and equity considerations particularly as it pertains to digital access, and worries that inaccurate and bias outputs may reinforce misconceptions rather than interrogation [15].

In responding to these challenges, while some institutions are choosing to ban the use of ChatGPT [12, 19], many educators are, instead, making a case for incorporating AI into education and developing critical awareness of AI literacy as a way to empower students and teachers to use the tool effectively and ethically [4, 18]. This paper adds to this conversation by presenting a case of how ChatGPT can be incorporated into an assignment to foster critical thinking.

3   Research Methodology and Assignment Description

The study involves a new cohort of Information Systems post-graduate students that enrolled at the start of the 2023 academic year. Students were given an academic writing assignment with the aim of developing and assessing their understanding and skills of academic writing, referencing, formatting, plagiarism and referencing. While the academic writing exercise has been a part of the course for well over 10 years, this is the first time that students have been asked to critically evaluate a response produced by a generative AI. The assignment required students to pose a question to ChatGPT in the domain of information systems project management and critically evaluate the response drawing on academic literature to support their critique. Some examples of prompts that students posed include: Can a project still be considered successful without satisfying the triple constraints of project management? Are there common leadership principles associated with failure for emerging project managers? Why Artificial Intelligence can help project managers and not take over their jobs. The assignment is part of their normal course requirements and contributes 2% towards their overall grade. A total of 84 assignments were submitted.

It is necessary to note that some of the students have little or no higher education background, having been admitted through a 'recognition of prior learning' system, which recognizes their knowledge and experience in the field of information systems; others have recent or relatively dated academic backgrounds from non-IS fields e.g. humanities or other disciplinary backgrounds.

The students were informed that, besides being used for the normal evaluation purposes, their assignments would also be used as a basis for this research project. They were allowed to opt out by a simple prefacing statement in their assignment (none did) and their plagiarism declaration also included the voluntary release of their assignment with the Creative Commons license. The assignments were anonymized by one of the team prior to analysis by the other team members.

Table 1 details the 4 components of the marking rubric as well as the correlations of the student marks between the rubrics. The correlations are fairly low i.e. the different rubric components measure quite different aspects of the student competencies.

Table 1. Marking rubric and correlations between rubric components.

Correlations	Academic writing style	APA referencing	Structure	Content
Academic writing style	1.000	0.501	0.548	0.595
APA referencing	0.501	1.000	0.345	0.345
Structure	0.548	0.345	1.000	0.417
Content*	**0.595**	**0.345**	**0.417**	**1.000**

* "Overall impression of the assignment (originality, critical evaluation, content)"

The research method adopted for this paper consists of two phases. To address the first research question – student perceptions of ChatGPT as a critical thinking tool – an analysis of the student actual answers was done. In particular, a thematic analysis of the critical evaluation by the student as found in their papers was done. In a somewhat circular approach, the themes uncovered were then compared to the type of themes which ChatGPT itself suggested could be used to evaluate itself. To address the second research question – to investigate how ChatGPT performs in generating a rubric and evaluating student essays – again ChatGPT was used directly, as detailed in Sect. 4.2.

4   Findings

4.1   What Were the Students' Perceptions on the Usefulness of ChatGPT's Outputs?

The conclusion of the assignment asks students' to reflect on their perceptions of using ChatGPT in the assignment. Most of the students conclude that the answers generated by ChatGPT are generally accurate, but there are some areas which are lacking.

A thematic analysis produced a number of recurrent themes in the student answers, as detailed in Table 2, column 2 (see infra). A more detailed discussion of some of the key themes with supporting student quotes includes the following.

- **Lack of depth.** Some students noted that the ChatGPT produced responses lack depth when compared to academic literature. One student, in particular, notes that ChatGPT is suitable for summarising information, but inadequate for solving problems,

It becomes obvious that ChatGPT is a natural language processing model which is focused on mimicking human conversation and not solving problems. When asked complex questions, ChatGPT responses focuses on a single factor, breaks down into rudimentary steps and presents it as a response. When the question is refined it paraphrases its original response to apply to the new refined question. ChatGPT may be suitable for summarizing information, but it has a long way to go in its development to be able to respond to broad and complex answers.

...the response was concise and factual. It gave several reasons, an explanation of each and a summary of all the reasons. However, academic literature indicated factors not included in the ChatGPT response...

While answers are correct, if you do not have adequate knowledge of a subject or do not perform proper research, you could easily be misled and only receive half the information.

For the most part, its reasoning and responses were in line with concrete evidence and were reliable to an extent. Although, a crucial flaw in the ChatGPT's response was that it excluded key information, however, this will not have a detrimental effect on the result.

- **Questionable accuracy.** Some students noted errors and emissions in the ChatGPT produced responses.

Overall, ChatGPT's response to a project management topic is knowledgeable and gives a thorough overview of the subject. It does, however, have several shortcomings, such as errors and omissions, and a lack of detail.

ChatGPT can give an answer to a question, but the level of its accuracy is still to be improved. Not all answers were as expected or as according to academic literature. The answers on whether a project manager or a scrum master are generic and not more specific and lacks context in explanation. As a result, there is some room for improvement on the level of detail ChatGPT gives when answering complex questions.

- **Bias in responses.** Various students demonstrated awareness of biases in the ChatGPT produced responses.

Based on the analysis done above it can be concluded that the scholar's experiences/research of the failure divide evident between industrialized and developing countries is still embedded in the DNA of ChatGPT's training data, leading to the AI tool leaning towards the same judgement in answering the main question at hand.

The answers provided by ChatGPT seems bias to making Project managers relevant in the modern Agile world. This makes one suspicious that ChatGPT uses sources that are from one side of the fence only.

- **Limitations with academic writing conventions.** Some students also commented that ChatGPT generated responses have limitations with regards to the style and convention of academic writing.

...it is certainly not yet at a level where it could adequately perform academic writing on behalf of humans as it is trained to mimic observed human output, but not necessarily trained to understand it. Caution, supervision and scrutiny should be exercised when seeking to use this tool as a writing aid; especially in an academic or professional setting.

The general writing of ChatGPT also seems to have particular writing style and I think it does not hold up to the expectation or standard for writing academic papers, essays, articles and so on.

...it does not contain citations, references or original source of the information returned.

- **A need for human expertise.** Many of the students conclude that while ChatGPT can provide information and insights, it cannot replace human judgement, which is needed to make informed decisions in different contexts.

In summary, Chat GPT has demonstrated its ability to provide information and insights, but it should be used in conjunction with human expertise to make informed decisions in the rapidly evolving and dynamic landscape of project management.

As a result, while ChatGPT can be a useful tool for project management, it should be used with caution and in conjunction with other tools, and human project managers are ultimately responsible for a project's success.

It is worth noting that this is a computer and therefore requires a human mind to assist it, it does not think but only gives information according to the data it has been trained on.

- **Lack of context specificity or localisation.** One major criticism was that the Chat-GPT answers were 'general' i.e. applying to a global context, not the specific South African situation, even when specifically requested to.

The answer of OPENAI (2023) does not contain mention of diversity in project teams nor the importance of applying diversity management in projects. [...] Diversity management and project based learning was not included in the Chat-GPT response to the question where Hewitt (2008) and Fioravanti et al. (2018) illustrates the importance of these social factors in project management.

However, while Chat GPT can provide a comprehensive comparison of these approaches, it is not able to provide personalized recommendations for specific projects. Its responses are based on academic literature and lack the context of real-world scenarios.

The answers on whether a project manager or a scrum master are generic and not more specific and lacks context in explanation.

Besides the thematic analysis, we thought it might also be interesting to see to what extent ChatGPT would be able to generate "self-reflectively" the type of typical criticisms leveled against itself. Note specifically that the authors do *not* mean or intend hereby to attribute any agency or introspective capabilities to ChatGPT; this is merely an 'exercise' to see how "creative" ChatGPT generates possible themes. In a way, this can be seen as an attempt to see how well ChatGPT could perform as a research assistant tool. So the prompt posed to ChatGPT was: *"If I ask students to critically evaluate the quality of ChatGPT responses in the domain of project management, which themes should I be looking for in their evaluation?"* Table 2 below shows the themes/topics suggested in its first (1) and/or second (2) response and mapped onto the themes that the authors originally identified.

Table 2. Themes mentioned by students compared with ChatGPT suggestions.

	Theme mentioned in at least 2 student essays?	Criterion suggested by ChatGPT?
Accuracy/Reliabilty	Yes	Yes (1) (2)
Completeness	Yes	Yes (1) (2)
Clarity	**No**	Yes (1) (2)
Relevance	Yes	Yes (1) (2)
Depth	Yes	Yes (1)
Bias	Yes	Yes (1)
Engagement	**No**	Yes (1)
Originality	Yes	Yes (2)
Organisation	**No**	Yes (2)
Currency	Yes	Yes (2)
Context specificity/Localisation	Yes	**No**
Lacking academic writing style	Yes	**No**

Interestingly enough, only *two themes* were not mentioned by ChatGPT, namely that of ChatGPT lacking context specificity i.e. not localising or particularization its answer to a specific situation, organisation or regional context; and it not formatting its answer sufficiently in an academic writing style and/or conventions (esp. Lacking references, but also other stylistic limitations). All ten themes suggested by ChatGPT have face validity and *could* have occurred in the empirical data set. As it was, only three of the ten were *not* found or identified by (at least two) students: clarity of expression e.g. lack of using jargon, engagement i.e. keeping the reader interested, and organisation or structure of the answer (but note that the assignment had a prescribed structure). This hints towards the possible use of ChatGPT in qualitative research in checking for possible themes in specific research data sets. This aspect – i.e. using ChatGPT as a research assistant tool – is worthy of future research.

4.2   How Did ChatGPT Perform in Producing a Rubric and Evaluating Students' Assignments?

In exploring the level of critical thinking displayed by the students in the assignments, we began by asking ChatGPT to generate a scoring rubric to assess critical thinking. With different prompts and attempts on different occasions, different rubrics were produced. Table 3 and 4 present two different rubrics. In Table 4, while most of the criterias are expected (e.g. clarity of thought, evidence, use of reasoning and counterarguments), one criteria stands out as uncommon for a rubric for assessing critical thinking – creativity. In the descriptors, creativity is described as demonstrating "creativity and originality, providing a unique perspective on the topic". Given that academic writing mostly fore-grounds logic and reasoning, 'creativity' is an unusual criteria in this context. Table 3 displays another unusual criteria "development of solutions and recommendations". In the level descriptor for the category 'Excellent', this criteria is described as "Solutions or recommendations are effectively developed and justified, taking into account poten-tial consequences and ethical implications". While potential consequences and ethical implications are features that need to be considered in research writing, it is not a usual requirement for student essays.

Another feature that is worth noting is the level descriptors of the level 'Emerging' in Table 4. In common English language, 'emerging' suggests development. Something which emerges is something that is starting to grow. However, for the criteria 'clarity of thought' the level descriptor for the level 'emerging' is outlined as "The ideas presented are unclear, poorly organised, or illogical". The title 'Emerging', which carries a positive connotation, as such, clashes with the descriptor, which carries a negative connotation.

After creating various ChatGPT produced rubric, we chose one (the rubric outlined in Table 4) and asked ChatGPT to evaluate the students' assignments using its own rubric. To test the reliability of the results, we asked ChatGPT to evaluate a given assignment using the same rubric more than once. These results were then compared to that of a human marker (one of the researchers marked selected essays using the same ChatGPT produced rubric). As can be seen in Fig. 1 below, different attempts with ChatGPT on different occasions produced different results. In particular, the important variations in results surfaced in two categories: creativity and use of counterargument. It can be argued that while a variation between "Exemplary (4)" and "Proficient (3)" is subject to debate, whereas a difference between "Proficient (3)" and "Developing" (2) is quite significant. "Developing" is a category that can be described as 'unsatisfactory', as it shows a lack of some sort, while "Proficient" is at the level of satisfactory. A discrepancy between these two scores thus can translate to a significant gap in ascertaining the quality of the work.

Table 3. A critical thinking rubric produced by ChatGPT (B)

Criterion	Excellent (4)	Good (3)	Fair (2)	Poor (1)
Identification and Analysis of the Problem	Problem is clearly and accurately identified, and analyzed in depth	Problem is identified clearly, but analysis is somewhat shallow or incomplete	Problem is identified, but analysis is superficial or incomplete	Problem is not identified or analysis is completely inadequate
Use of Evidence and Reasoning	Relevant evidence is effectively used to support claims, and arguments are clear, coherent, and logical	Relevant evidence is used to support claims, but arguments are somewhat unclear, incoherent, or illogical	Relevant evidence is used, but arguments are often unclear, incoherent, or illogical	Evidence is not used effectively, or arguments are completely unclear, incoherent, or illogical
Evaluation of Sources	Sources are effectively evaluated for credibility and reliability, and are cited appropriately and ethically	Sources are generally evaluated for credibility and reliability, but may not be cited appropriately or ethically	Sources are evaluated, but not effectively or consistently, and may not be cited appropriately or ethically	Sources are not effectively evaluated or cited appropriately or ethically
Consideration of Context and Assumptions	Broader context is effectively considered, and assumptions are identified and critically examined	Broader context is generally considered, and assumptions are identified, but not always critically examined	Broader context is considered, but not consistently or effectively, and assumptions may not be identified or critically examined	Broader context is not considered, or assumptions are not identified or critically examined
Development of Solutions and Recommendations	Solutions or recommendations are effectively developed and justified, taking into account potential consequences and ethical implications	Solutions or recommendations are developed and justified, but may not fully take into account potential consequences and ethical implications	Solutions or recommendations are developed, but may not be fully justified or take into account potential consequences and ethical implications	Solutions or recommendations are not effectively developed or justified, or do not take into account potential consequences and ethical implications

Table 4. A critical thinking rubric produced by ChatGPT (A)

Criterion	Exemplary (4)	Proficient (3)	Developing (2)	Emerging (1)
Clarity of Thought	The ideas presented are clear, well-organized, and logically developed, making it easy to follow the argument	The ideas presented are generally clear and logically developed, but could be better organized	The ideas presented are somewhat clear, but lack logical development or are not well-organized	The ideas presented are unclear, poorly organized, or illogical
Depth of Analysis	The ideas presented demonstrate a deep level of thinking, insight, and analysis, showing a thorough understanding of the topic	The ideas presented demonstrate a good level of thinking and analysis, showing a solid understanding of the topic	The ideas presented demonstrate some level of thinking and analysis, but may lack depth or complexity	The ideas presented lack sufficient analysis or are overly simplistic
Evidence	The ideas presented are supported by strong and relevant evidence from credible sources	The ideas presented are generally supported by relevant evidence, but may have some gaps or lack credibility	The ideas presented have some supporting evidence, but may lack relevance or credibility	The ideas presented lack sufficient or credible evidence to support the argument
Creativity	The ideas presented demonstrate a high level of creativity and originality, providing a unique perspective on the topic	The ideas presented are somewhat creative and offer some original insights into the topic	The ideas presented lack creativity or originality, and may be too similar to existing ideas	The ideas presented lack creativity and originality, and rely heavily on existing ideas
Relevance	The ideas presented are highly relevant to the assignment topic and demonstrate a clear understanding of the assignment requirements	The ideas presented are generally relevant to the assignment topic, but may have some tangential or off-topic elements	The ideas presented lack relevance to the assignment topic, or demonstrate a limited understanding of the assignment requirements	The ideas presented are completely irrelevant to the assignment topic, or demonstrate a complete misunderstanding of the assignment requirements
Use of Reasoning	The ideas presented demonstrate a strong use of reasoning and logical thinking, avoiding fallacies and emotional appeals	The ideas presented generally demonstrate sound reasoning and logical thinking, but may have some logical fallacies or emotional appeals	The ideas presented show some use of reasoning and logical thinking, but may have significant logical fallacies or emotional appeals	The ideas presented lack sound reasoning or logical thinking, relying heavily on fallacies or emotional appeals

(*continued*)

Table 4. (*continued*)

Criterion	Exemplary (4)	Proficient (3)	Developing (2)	Emerging (1)
Use of Counterargument	The ideas presented address potential counterarguments and demonstrate an awareness of alternative perspectives	The ideas presented generally address potential counterarguments, but may miss some or lack depth	The ideas presented fail to address potential counterarguments, or demonstrate a lack of awareness of alternative perspectives	The ideas presented actively avoid addressing potential counterarguments, or demonstrate a complete lack of awareness of alternative perspectives
Critical Evaluation	The ideas presented demonstrate a critical evaluation of the topic, rather than simply accepting it at face value	The ideas presented generally demonstrate some level of critical evaluation of the topic, but may have some gaps or lack depth	The ideas presented lack critical evaluation of the topic, and may simply accept it at face value	The ideas presented demonstrate a complete lack of critical evaluation

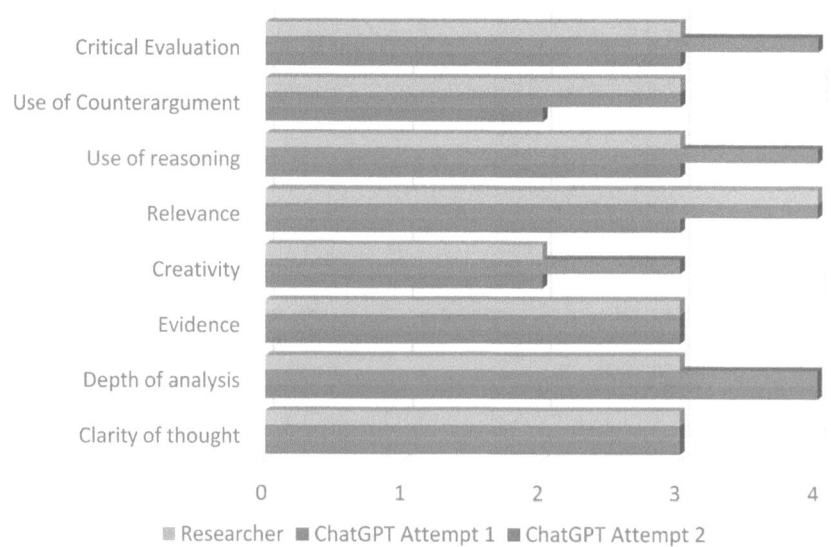

Fig. 1. Comparison of scores: Researcher versus ChatGPT.

5 Discussion and Conclusions

Critical thinking is an important skill to develop during higher education studies. Developing AI literacy, as a subset of critical thinking, is an even more urgent need. This paper explored the potential of using ChatGPT generated assessment for both developing critical thinking and critical AI literacy in students. Students' perceived the ChatGPT produced responses as generally accurate, although they tended to lack depth, with some key information omitted, produced bias responses and had limitations with academic writing conventions. Overall, the students concluded that human expertise and judgement was needed in working with ChatGPT.

An interesting possibility emerged when we compared the themes that were identified during the qualitative analysis of the student answers, with the criteria that ChatGPT *itself* suggested should be used to evaluate its own answers critically. It generated ten criteria, of which seven were found in the student answers. Conversely, only two themes empirically identified in the student answers were *not* mentioned by ChatGPT, namely ChatGPT's lack of context specificity and its inability to fully compose its answers in the required academic style. This hints towards a possible constructive use of ChatGPT in supplementing qualitative analysis.

Despite some suggestions that ChatGPT can be used to generate rubrics and grade assignments [7, 10], our findings indicate that ChatGPT produced rubrics cannot be taken as is, but these need to be reviewed to determine if the criteria, level indicators and descriptors are indeed fitting for the occasion. Critical thinking means different things to different people, from a limited view of criticality linked to cognition and rational thought, to more dialogical, creative, even caring approach to critical thinking [14]. However, as ChatGPT does not yet reveal its sources, it is difficult to understand where certain criteria come from, what theories they might be framed with, unless one engages ChatGPT in conversation and dialogue. Furthermore, with inconsistent results, the reliability of using the tool as a grading device is also questionable.

ChatGTP might be a useful thinking, learning and teaching companion but needs conversation, engagement, a lecturer's own critical thinking and analysis of ChatGPT recommendations, to be used in meaningful ways in learning and teaching. However, seeing how rapidly the large language model is developing, we might find that some of the current limitations will be resolved soon. What we believe is important, is that students understand both the potential and the limitations of these tools to support their learning. As such, critical AI literacy for both lecturers and students might be one of the most important elements of any curriculum, and an exercise such as the one discussed in this paper, might be one way of opening up the conversation.

References

1. Baidoo-Anu, D., Owusu Ansah, L.: Education in the era of generative artificial intelligence (AI): understanding the potential benefits of ChatGPT in promoting teaching and learning. J. AI **7**(1), 52–62 (2023)
2. Bang, Y., et al,: A multitask, multilingual, multimodal evaluation of chatgpt on reasoning, hallucination, and interactivity. arXiv preprint arXiv:2302.04023 (2023)

3. Bommarito II, M., Katz, D.M.: GPT Takes the Bar Exam. arXiv preprint arXiv:2212.14402 (2022)

4. Cardon, P., Fleischmann, C., Aritz, J., Logemann, M., Heidewald, J.: The challenges and opportunities of AI-assisted writing: developing AI literacy for the AI age. Bus. Prof. Commun. Q. **86**(3), 257–295 (2023). https://doi.org/10.1177/23294906231176517

5. Castelli, P.R.: ChatGPT AI passes the medical licensing examination in the US. IBSA Foundation for Scientific Research (2023). https://www.ibsafoundation.org/en/blog/chatgpt-passes-medical-licensing-examination

6. Coleman, M.: Using ChatGPT in higher education, Honours Thesis, University of Cape Town [Draft/Unpublished] (2023)

7. Cronje, J.: The implications of ChatGPT for assessment in higher education. Webinar for the Academy of Science of South African (ASSAF) (2023). http://hdl.handle.net/20.500.119 11/275

8. Ellis, B.: How ChatGPT can help with grading. Technotes. https://blog.tcea.org/chatgpt-gra ding/. Accessed 2 June 2023

9. Guo, B., et al.: How Close is ChatGPT to Human Experts? Comparison Corpus, Evaluation, and Detection. arXiv preprint arXiv:2301.07597 (2023)

10. Iskender, A.: Holy or unholy? Interview with open AI's ChatGPT. Eur. J. Tourism Res. **34**, 3414 (2023)

11. Kasneci, E., et al.: ChatGPT for good? On opportunities and challenges of large language models for education. Learn. Individ. Differ. **103**, 102274 (2023)

12. Mackay, T.: What is AI? Why was ChatGPT banned? 'Greatest cheating tool ever invented' says teacher. The Scotsman. https://www.scotsman.com/lifestyle/tech/what-is-ai-why-was-chatgpt-banned-greatest-cheating-tool-ever-invented-says-teacher-4007003. Accessed 2 June 2023

13. Nisar, S., Aslam, M. S.: Is ChatGPT a good tool for T&CM students in studying pharmacology? SSRN Electron. J. (2023). SSRN 4324310

14. Nomdo, G.: Unpacking the notion of 'criticality' in liberatory praxis: a critical pedagogy perspective. Crit. Issues Teach. Learn. (CriSTaL) **11**(SI), 50–70 (2023). https://www.cristal. ac.za/index.php/cristal/article/view/644

15. Rasul, T., et al.: The role of ChatGPT in higher education: benefits, challenges, and future research directions. J. Appl. Learn. Teach. **6**(1), 1–16 (2023)

16. Rosenblatt, K.: ChatGPT passes MBA exam given by a Wharton Professor. NBC News (2023). https://www.nbcnews.com/tech/tech-news/chatgpt-passes-mba-exam-wharton-professor-rcna67036

17. Rudolph, J., Tan, S., Tan, S.: ChatGPT: Bullshit spewer or the end of traditional assessments in higher education? J. Appl. Learn. Teach. **6**(1), 342–363 (2023)

18. Southworth, J., et al.: Developing a model for AI across the curriculum: transforming the higher education landscape via innovation in AI literacy. Comput. Educ.: Artif. Intell. **4**, 100127 (2023)

19. Tlili, A., et al.: What if the devil is my guardian angel: ChatGPT as a case study of using chatbots in education. Smart Learn. Environ. **10**(1), 15 (2023)

20. Yeadon, W., Inyang, O.O., Mizouri, A., Peach, A., Testrow, C.P.: The death of the short-form physics essay in the coming AI revolution. Phys. Educ. **58**(3), 035027 (2023)

A Consolidated Catalogue of Question Types for Programming Courses

Anitta Thomas(✉) (iD)

School of Computing, The Science Campus, University of South Africa, Florida Park, South
Africa
thomaa@unisa.ac.za

Abstract. Formulating questions, whether for assessments or to create exercises,
is an essential component of teaching programming. The development of well-
constructed questions requires considerable effort. In this research, a consolidated
catalogue of question types was developed to assist programming instructors with
the formulation of questions. Existing classifications of question types in both
programming and computing were consulted in developing the catalogue. The
catalogue was constructed using a systematised literature review by referring to
the literature and unifying different categories of question types described in the
literature. The catalogue consists of fourteen base types of questions, with the possibility of variations within most base types. The catalogue is primarily intended
to assist instructors in designing and formulating more diverse types of questions
in programming courses. Additionally, the author envisages that this catalogue
can serve as a starting point for discussions on how to map current educational
taxonomies to question types, the different skills related to these questions, appropriate formats (for instance, multiple choice vs open format questions) and the
appropriate qualification levels at which these questions can be presented.

Keywords: Question types · programming courses · programming assessments ·
classification of questions

1 Introduction

Two general categories can be used to classify computing education research: one focuses
on learning, and the other on teaching computing subjects. Learning processes and
learners' challenges in computing subjects are some of the themes of interest in the first
category. The development of effective teaching strategies and assessments is of interest
in the second category of research, which is focused on instructors [1]. This study, which
specifically focuses on computer programming, falls into the second category.

Teaching programming entails formulating questions for learners at some point during the delivery of a course, just like teaching any other computing subject. These questions are typically used for in-class activities [2, 3], laboratory exercises [4], formative
assessments [5] and summative assessments [6]. The development of questions requires
significant effort from the instructors [7], especially when such questions are carefully

© The Author(s), under exclusive license to Springer Nature Switzerland AG 2024
H. E. Van Rensburg et al. (Eds.): SACLA 2023, CCIS 1862, pp. 119–133, 2024.
https://doi.org/10.1007/978-3-031-48536-7_9

designed to develop different skills in the learners [1], to accommodate the diverse abilities and interests of learners [8], to support autograding [9], to reduce dishonesty [7] and to allow for specific formats (for example, multiple-choice questions (MCQs)) [10]. Any programming instructor would agree that formulating well-designed questions of diverse types for each course delivery can be a challenging and time-consuming task.

Given that developing novel questions requires much effort, programming instructors can benefit from an in-depth catalogue of question types that could be used as the basis for designing appropriate questions. Such a catalogue should be able to answer the question, "What types of questions can be asked in programming courses?". It could also be used, for example, by novice instructors to formulate new questions or experienced instructors wishing to introduce more diversity into their existing question pools. Learners benefit from diverse types of questions; as such, diversity supports in-depth learning and improves their learning experience [1]. Additionally, such a catalogue could be used to create balanced repositories of programming questions, which would be of great value to instructors [11].

Several catalogues of question types used in computing subjects can be found in the existing literature [1, 4, 6, 12–14]. In [1] and [4], twelve unique question types are presented for computing subjects. Eight types of questions considered a set of eight unique skills typically used in a programming exam are described in [6]. In [12], ten different question types for computing are discussed. [13] and [14] present twelve and fourteen question types, respectively. These catalogues have been derived using different methods, including reviewing curricula, synthesising computing education literature [12], the experiences of instructors [4, 12], analysing existing assessments [6, 14], and examining textbook exercises [13, 14]. A comparison of these catalogues shows that they share similarities and differences in their content. Additionally, some are specifically proposed for programming [6, 13, 14], and some for computing in general [1, 4, 12], which could be applied to programming. An instructor referring to literature would thus face the challenge of analysing several different catalogues of question types for programming subjects.

This research aims to create a consolidated catalogue of question types that can be used in programming courses to address the above-mentioned challenges. The catalogue is intended as a resource to assist programming instructors in developing diverse and valid question types to enhance their teaching as well as their learners' learning experience. A systematised literature review [15] was conducted to identify the question types mentioned in the literature in support of the aim of the present research. The catalogue resulting from the synthesis of the existing literature includes a list of question types that can be used in a programming course.

The primary goal of this research is to develop an in-depth catalogue of question types to support instructors in designing questions for programming courses. The main objective is achieved by identifying and synthesising existing question types from the literature to develop a catalogue of relevant question types for programming courses. The identification of existing question types is achieved using a systematised literature review [15].

This paper is structured into five sections. Section 1 contains an introduction to the paper. Section 2 includes a discussion of the research method, specifically discussing the systematised literature review in general, including the details of the review conducted for this research. The developed catalogue of question types is presented in Sect. 3. Each type of question is discussed in detail. Section 4 comprises a general discussion. Section 5 presents the conclusion, along with some uses of the catalogue and suggestions for future work.

2 Research Method

This section briefly explains our research method. A systematised literature review (SLR) was conducted in this research to support the objective of developing a catalogue of question types that can be used in programming courses. A systematised literature review is a subtype of a systematic literature review that is limited in scope [15]. In this paper, the acronym SLR is used for the term "systematised literature review", not for "systematic literature review".

An SLR falls short of a systematic literature review in several aspects, of which the first aspect is that an SLR permits restricted search space for finding relevant publications. For example, a literature search can be limited to certain databases as well as years of publication. The second aspect is that an SLR endorses a lenient assessment of how articles are selected or excluded for analysis in the review. The third aspect is that an SLR supports the review of literature by a single reviewer whereas a systematic literature review aims for a comprehensive review that considers all relevant publications and utilises more than one reviewer to conduct the review [15].

Due to its limited scope, an SLR cannot claim comprehensiveness. However, an SLR still employs a transparent approach to reviewing literature. The goals guiding the review are stated explicitly, along with the databases used. The inclusion and exclusion criteria can also be clearly stated to justify the articles included in the review. By making aspects of an SLR transparent, the replicability of such a review is improved.

The literature review was conducted to identify and synthesise existing question types from relevant publications. The review started with a literature search to identify articles that discussed different question types for programming. To wit, the literature search was guided by the question, *"What types of questions have been used in programming courses?"*.

Four online databases were used for the literature search: *Web of Science, IEEE Xplore, ACM Digital Library* and *Springer Link*. These databases were selected based on the expectation that most of the computing literature can be found in them. The search started with the aim of finding publications that focused on classifications of question types in programming and computing. Question types in computing were also considered based on the premise that numerous types in computing would be relevant to the programming subfield. The keywords, including taxonomy, classification, catalogue, programming, computing, question types and task types were used to formulate search strings. The initial search retrieved only six articles that focused on the classification of question types. To include more articles, it was decided to include articles that discussed question types and questions for other purposes and were not necessarily focused on classifications of these question types.

The search for relevant literature was limited to research published between 2002 and 2022 (the past 20 years). To identify other relevant papers, reference lists of selected publications were used in addition to these four databases. In this study, only peer-reviewed publications in English were considered. If there were more than one publication with the same content, only the latest publication was considered for review. An article had to explain several question types, either with or without examples, to be considered for analysis.

Following the criteria discussed above, twenty-nine articles were chosen for analysis[1]. The catalogue in this research was developed by reading each of these twenty-nine articles individually and recording the different types of questions that are mentioned in them. The analysis process started with articles that focused on classifications of question types so that these different classifications could be amalgamated into a single catalogue. The rest of the articles were then analysed to confirm or add new question types to the catalogue.

3   A Catalogue of Question Types for Programming

This section presents the analysis and synthesis of the twenty-nine articles mentioned in Sect. 2. The synthesis is presented as a catalogue with fourteen different base types of questions. A general description, what is included in a typical question, what is expected in the answer, a simple example question, and a discussion that includes possible variations of each question base type are provided. All relevant examples use *Java* as the programming language, which was an arbitrary choice. The order of the presentation of the question base types bears no significance. In presenting each question base type, the focus is on the type itself, with no mention of the format (such as MCQs or free-form) in which it may be asked.

3.1   Question Type 1: Write code

Type description: Learners are expected to write code for a problem or a scenario in this type of question [4–6, 9, 11–14, 16–29].
Question details: The question describes the problem to be solved. This type of question can be stated in the text [4], but can also utilise diagrams [14] (for example, flowcharts and UML diagrams [13]). It may state specific requirements to be satisfied in the solution. Examples of requirements are the need for specific input and/or output [16], to achieve a certain level of efficiency [4], and to use specific programming constructs [9] or concepts such as recursion and inheritance.
Answer details: The expected answer is code. Depending on the question, answers can be complete programs or parts of programs, such as a function or a class.
Example: Write a function that can generate and return a set of six unique random numbers between one (1) and 52 to simulate *quick play* in the national lottery scheme.
Discussion: This type of question is well-known and frequently used in programming courses. One can formulate questions with all the requirements clearly discussed in the question. That said, this type also allows open-ended questions, where learners must exert themselves to understand the problem.

---

[1] List of articles and search string.

3.2 Question Type 2: Write Code by Using Specific Code

Type description: In this type of question, learners are expected to write code by using specific code [4–6, 9, 11, 13, 26, 30].

Question details: The details of the question for this type are similar to those of the question for Type 1. Additionally, these questions include code or references to code that must be utilised in the answer.

Answer details: The details of the answer for this type are similar to those of the answer for Type 1.

Example: Write a program that accepts a positive integer representing the number of entries a player wishes to play in the national lottery scheme. The program must then generate unique sets of six numbers between one (1) and 52. The number of sets is equivalent to the positive integer specified by the user. The function that you wrote as the response to the example question given under Question Type 1 must be used in this program.

Discussion: Such questions can be of limited scope. Further, in general, the requirements to be satisfied by the answer are clearly stated in such a question. The code that must be used in the solution could be provided in a variety of ways. It might be a file attached to the question, a component of the question itself, or something the learners have already completed.

This type of question can be used for writing code to test given code as well [6, 9, 11, 13].

3.3 Question Type 3: Modify Specific Code

Type description: Learners are expected to produce code by modifying given code [2, 4, 5, 9, 11–14, 16–19, 21, 24, 26, 28, 30, 31]. Different forms of code modification are possible. Examples of modifications are the completion [4, 16], extension [13, 14], and transformation of given code [4, 32] to satisfy specific requirements.

When comparing this type to Type 2, this type is focused on modifying while Type 2 is focused on reusing specific code.

Question details: The details of the question for this type are similar to those of the question for Type 1. Additionally, these questions include code that must be modified to produce the solution.

Answer details: The expected answer is code. Depending on the question, answers can be complete programs or parts of programs, such as a function or a class.

Example: The code fragment given below counts and outputs the number of odd numbers between one (1) and 100 (inclusive). Modify the code fragment to count the number of even numbers by using a for loop instead of a while loop.

```
int count = 0;
int num = 1;
while(num <= 100){
    if((num % 2) != 0){
        count++;
    }
    num++;
}
System.out.println(count);
```

Discussion: Such questions are often limited in scope. In general, the requirements to be satisfied by the answer are clearly stated in such a question. Examples of requirements are optimising the code [13, 14], adding a certain functionality to the code [5, 19], changing the functionality of the code [28] and using different programming constructs in the provided code [13].

When the question is to complete a given code, the lines of code that need to be completed could be anywhere in the given code. An extension of the given code could be to extend the functionality of the code. A transformation could include rewriting given code in a different programming language [4] or using different programming constructs [13].

3.4   Question Type 4: Create a Design

Type description: Learners are expected to provide a design as the solution to the stated problem [6, 11, 12, 22, 25].

In this type, learners are not expected to provide an implementation, as is the case with Type 1.

Question details: A question of this type will provide a general description of the problem and the solution format [6, 11, 12, 22, 25].

Answer details: The answer to such a question would be a design.

Example: Draw a flowchart that depicts a possible solution for the function stated in the example in question Type 1.

Discussion: Examples of design formats are diagrams [22] and pseudocode [25]. A system design created from a set of requirements can also be a valid question for this type.

3.5   Question Type 5: Debug Code

Type description: Learners are expected to identify errors in the given code and/or potentially explain and fix these errors [5, 6, 9, 11–13, 20, 22, 26, 27, 31–33].

Question details: A question of this type would contain a description as well as code to be examined by the learners. It could also state what type of error the code contains.

Answer details: The answer to this question varies depending on the question. It could be a list of lines of code containing errors, explanations of why errors occur [5, 13] or ways in which errors can be rectified [9].

Example: Study the class `Point` below:

```
public Class Point {                              //line 1
    private int x, y;                             //line 2
    public Point(){                               //line 3
        x = 0;                                    //line 4
        y = 0;                                    //line 5
    }                                             //line 6
    public Point(int newX, int newY){             //line 7
        newX = x;                                 //line 8
        newY = y;                                 //line 9
    }                                             //line 10
}                                                 //line 11
```

The code above has some syntactical and logical errors. Indicate the lines in which these errors occur and provide corrections for them.

Discussion: The errors in the code could be syntactical [12], logical [31] or run-time errors [5]. It is also possible to include code that combines syntactical, logical or run-time errors. A variation is to provide pseudocode for debugging [16].

3.6 Question Type 6: Analyse code

Type description: Learners are expected to analyse the given code [2, 4–6, 9, 11–14, 16–18, 20–22, 24, 27, 29, 31, 33–35].

Question details: A question of this type would have code in it and could ask for a trace table [13], the output of the code when executed [12, 17, 34], an explanation of the purpose of the code [5, 11, 20] or a description of the type of problem that could be solved using the given code [13, 14]. Irrespective of what the question specifies, to answer the question, the given code needs to be analysed.

Answer details: The answer to this question varies depending on the question. It could be a trace table [13], output [12, 17, 34], values of certain variables after execution [16], input values to generate specific output [4] or an explanation relating to the given code [5, 11, 20].

Example: Refer to the code below to answer the questions that follow:

```
int count = 0;
int num = 1;
while(num <= 10){
    if((num % 2) != 0){
        count++;
    }
    num++;
}
System.out.println(count);
```

(a) What is the output when the code is executed?
(b) Explain in one sentence what the code achieves.
(c) How many times does the while loop get executed?

Discussion: This is a well-known and frequently used question in programming. There are numerous possibilities with this type of question, as stated under the details of the question and answer. Additionally, learners can analyse code to establish efficiency [4, 11, 17], correctness [2, 4, 12, 14, 16, 17, 29], to examine the coding style and determine the effects of altering lines in the given code [4].

Another variation is to provide pseudocode or diagrams of data structures for analysis [17, 24].

3.7   Question Type 7: Write Preconditions for Specific Code

Type description: Learners are expected to write preconditions for specific code [13, 14].

Question details: A question of this type would contain code. It can also specify the format in which the preconditions must be written.

Answer details: The answer to the question will be preconditions.

Example: For the class given below, write pre- and postconditions for the method getX() using @pre and @post tags that can be included in the doc comments for this method.

```
public class Point {
    private int x, y;
    public Point(int newX, int newY){
        newX = x;
        newY = y;
    }
    public int getX(){
        return x;
    }
}
```

Discussion: A question of this type can ask for preconditions and/or postconditions for single, multiple blocks of code or scenarios. This type of question also requires learners to analyse and understand code, but it is listed separately because this type is different from typical code analysis tasks (see Sect. 3.6).

3.8   Question Type 8: Create a Model

Type description: Learners are expected to generate a model in this type of question [2, 12–14, 24]. The model could be generated based on specific code [2] or based on some programming concept [12].

When comparing Types 4 and 8, the focus of the former is on creating a solution to a stated problem, while the latter is focused on the abstraction of some code or concept.

Question details: The question describes what should be modelled and how the model should be represented [2, 12–14, 24]. A question of this type can also contain code [2] if the model is based on the code.

Answer details: Often, the expected model is represented using a diagram. Examples of such diagrams include UML and flowcharts [2].

Example: Based on the code snippets below, create a UML class diagram that shows classes and relationships between classes:

```
public class Point{
    private int x, y;
    //constructors and methods go here
}
public class Line{
    private Point endOne, endTwo;
    public Line(Point one, Point two){
        endOne = one;
        endTwo = two;
    }
    //constructors and methods go here
}
```

Discussion: The model representation can use standard (for example, UML) and non-standard visual languages (for example, a graph representation).

3.9   Question Type 9: Create an Example

Type description: Learners are expected to generate an example of some programming concept in this type of question [12, 24, 31].
Question details: A question of this type would provide details on the programming aspect of which the learner needs to provide an example. The question also specifies the format in which the example should be given [12, 24, 31].
Answer details: Depending on the question, the answer could be a sample code [24], a design [31] or even an explanation of a problem or scenario [12].
Example: Create a code fragment that demonstrates polymorphism using your own class hierarchy.
Discussion: One can formulate questions that require learners to generate examples in different formats.

3.10   Question Type 10: Theory Questions

Type description: Learners are expected to explain programming concepts they have encountered in the course [5, 11, 12, 19, 21, 24, 26, 27, 31, 32, 35].
Question details: A question of this type is formulated to produce an explanatory answer from the learners.
Answer details: The answer will be an explanation of the required concept.
Example: Explain the purpose of a destructor in a *Java* class.
Discussion: Questions can also be formulated to compare concepts [11] or justify a design or code choice [32] made by the learners. In general, these questions can be used to promote or establish learners' theoretical knowledge of programming.

3.11   Question Type 11: Focused on the Constructs of the Programming Language

Type description: Learners are expected to demonstrate an understanding of the constructs of the programming language used in the course [2, 9, 11, 16, 21, 24, 29–31, 33].

Question details: Questions of this type can have a textual description [33], code fragments [31], designs [11] and output [16].

Answer details: The answer can be a description or a list of programming constructs.

Example: List the *Java* classes in the *AWT* package that can be used to render a text area, a scroll bar and a button.

Discussion: The focus of this type of question is purely on the constructs of the programming language, while numerous other base types, for example, Types 1, 2 and 3 assume knowledge of programming language constructs to successfully answer them.

3.12   Question Type 12: Create a Question

Type description: Learners are expected to design questions for the listed topics [4, 34].

Question details: Questions of this type describe the concepts to be tested by the question [4]. It can also state the format in which the question should be presented [34].

Answer details: The answer will be the questions themselves.

Example: Create a question of Type 1 (write code) to demonstrate an implementation of recursion.

Discussion: Although this type of question might not be as commonly used as other types (for example, writing code), such an activity could enhance learners' learning experience [4, 34]. Because the question or parts of the question and response or parts of the response developed for this type can be viewed as examples, there may be a close relationship between this type and Type 9.

3.13   Question Type 13: Run the code in an IDE

Type description: Learners are expected to execute a given code in an integrated development environment (IDE) [14].

Question details: Questions of this type provide the code and, most likely, the IDE to run the code.

Answer details: The answer would be a demonstration that the code was executed in IDE, for example, a screenshot.

Example: Provide a screenshot of the output produced by running the code provided in the example in Question Type 3 in *BlueJ*. You will have to add additional code to make the given code produce an output.

Discussion: The purpose of this type of question is conceivably to determine whether the learner has installed and can use an IDE correctly. This type of question can be used to validate that the student has installed or set up any required development platforms for the successful completion of the programming course.

3.14 Question Type 14: Self-reflection

Type description: Learners are expected to reflect on some aspects that they have encountered in the course [12].
Question details: A question of this type would describe the aspect that needs to be reflected upon [12].
Answer details: The answer to such a question is the individual reflection of the specified aspect [12].
Example: Write a reflection on your learning experience after completing the mini project in this course, focusing on aspects you learnt the most about.
Discussion: This might be an unusual question for a programming course, but it provides an opportunity for learners to reflect on their learning of specific concepts or assessments. Instructors can also ask learners to self-evaluate their programming tasks or projects.

4 Discussion

The catalogue presented above includes different base types of questions, most with numerous variations, that can be used in a programming course. In general, the catalogue encompasses more practical than theoretical types of questions, presumably owing to the practical nature of programming. No claim to comprehensiveness can be made about the types of questions since only a limited number of resources were consulted to develop this catalogue. However, it can be said with some degree of certainty that most question base types that can be formulated in programming courses are covered in the catalogue based on the researcher's teaching experience.

The classification of questions was somewhat subjective, not unequivocal, since the boundaries of each base type are often 'fuzzy'. For example, consider Question Types 1, 2 and 3, which all deal with writing code. One can merge these three into one base type and classify them as variations of a single base type. It is also possible to have more base types of questions, especially if some of the variations listed within each base type are expanded into separate base types; that is, different classifications of these question types are possible.

As demonstrated in the examples in each question base type, one can formulate questions that fall strictly into one base type. However, one can also combine questions of different base types when formulating questions in a programming course [4]. Such combinations can assist in creating richer assessment options. For example, one can combine Question Types 1, 4 and 14 to create a mini-project assessment in a programming course. The types of questions that you choose for a programming course will depend on numerous aspects. Such aspects include whether you are developing questions for in-class or laboratory activities or hand-written examinations and the qualification level at which the programming course is offered. That said, most of these question types can be used for different qualification levels when appropriately scaled.

The developed catalogue in its current form is intended to provide instructors with a resource to locate different possibilities for formulating questions in programming courses. However, numerous aspects can be added to the consolidated catalogue to make it more meaningful for instructors. Examples of such aspects include the learning skill associated with each question type, the formats each question type can accommodate and the marking criteria for each question type. One could even consider a list of features that can be used to further describe the questions in each type. Examples of such features include question and answer formats, required tools, and the expected time duration to complete the answer successfully. There are numerous possibilities for further research to enrich the catalogue.

The SLR utilised five existing classifications [4, 6, 12–14], where [4] and [12] focused on question types in computing at large and [6, 13, 14] focused on question types in programming. When comparing the classification in [4] to the catalogue in this research, it is noted that [4] does not explicitly discuss questions of Base Types 7, 9, 10, 11, 13 and 14. All the question types mentioned in [4] are included in the developed catalogue, albeit using different base type titles and classifications. Similarly, the classification in [12] does not explicitly list Question Base Types 2, 3, 7, 11, 12 and 13 discussed in Sect. 3. Again, the classification titles in [12] are different from the classifications in this paper.

The classification in [6] is focused on examination questions in introductory programming courses. [6] excludes Question Base Types 7 to 14 from the developed catalogue. Understandably, not all question base types in Sect. 3 are suitable for an introductory programming hand-written examination. When comparing [14] to the developed catalogue, it does not discuss question Base Types 2, 9, 10, 11, 12 and 14. Similarly, [13] excludes Question Base Types 9, 10, 11, 12 and 14. Based on the comparisons, it can be concluded that the developed catalogue has certainly consolidated and expanded on the existing individual classifications considered in this research.

It should be noted that a limitation of this research is that it used an SLR instead of a complete systematic literature review. Only a limited number of databases and publications published during a specific period were considered for the literature review in this study. Moreover, the review was conducted by only one reviewer, i.e., the researcher.

5 Conclusion

In this research, a consolidated catalogue of fourteen question base types for programming courses was developed using a systematised literature review. Each base type is presented with sufficient detail, including variations for each sub type, so that both novice and experienced instructors can use the catalogue to improve their skills in formulating diverse questions in programming courses.

Apart from being used for formulating questions, the catalogue has other potential uses. The catalogue could be used as a starting point for discussions about question types and expanding the types of questions in programming. The catalogue could also be used as a basis for creating shared repositories of programming questions.

In its current state, the catalogue does not indicate certain aspects. First, it does not discuss types of skills, specifically connecting them to the relevant educational taxonomies required to answer each type of question. Second, the catalogue does not discuss in what formats (closed vs. open) questions of a base type can be formulated. Third, it would be worthwhile to establish at which qualification levels and course activities (such as in-class activities, laboratory exercises, assignments and examinations) each question base type would be appropriate. All these aspects could be considered for future research.

The developed catalogue offers numerous opportunities for extension as well as for future research. However, in its current state, it is meant to provide instructors with possibilities for formulating a variety of questions in their programming instruction.

Acknowledgements. The author would like to thank the anonymous reviewers for their contributions towards improving this paper. B. Chimbo is acknowledged for our discussions in the initial stages of this research. A.E. Botha is also acknowledged for providing suggestions to improve this paper.

References

1. Hazzan, O., Lapidot, T., Ragonis, N.: Introduction – what is this guide about? In: Hazzan, O., Lapidot, T., Ragonis, N. (eds.) Guide to Teaching Computer Science: An Activity-Based Approach, pp. 1–12. Springer London, London (2011). https://doi.org/10.1007/978-0-85729-443-2_1
2. Hauswirth, M., Adamoli, A.: Teaching Java programming with the Informa clicker system. Sci. Comput. Program. **78**, 499–520 (2013). https://doi.org/10.1016/j.scico.2011.06.006
3. Reckinger, S., Hughes, B.: Strategies for implementing in-class, active, programming assessments. In: Proceedings of the 51st ACM Technical Symposium on Computer Science Education, pp. 454–460. ACM, New York, NY, USA (2020). https://doi.org/10.1145/3328778.3366850
4. Ragonis, N.: Type of questions – the case of computer science. Olympiads Inform. **6**, 115–132 (2012)
5. Sorva, J., Sirkiä, T.: Embedded questions in ebooks on programming – useful for a) summative assessment, b) formative assessment, or c) something else? In: Proceedings of the 15th Koli Calling Conference on Computing Education Research, pp. 152–156. ACM, New York, NY, USA (2015). https://doi.org/10.1145/2828959.2828961
6. Sheard, J., et al.: Exploring programming assessment instruments: a classification scheme for examination questions. In: Proceedings of the Seventh International Workshop on Computing Education Research, pp. 33–38. ACM, New York, NY, USA (2011). https://doi.org/10.1145/2016911.2016920
7. Simon: Designing programming assignments to reduce the likelihood of cheating. In: Proceedings of the Nineteenth Australasian Computing Education Conference, pp. 42–47. ACM, New York, NY, USA (2017). https://doi.org/10.1145/3013499.3013507
8. Torrey, L.: Student interest and choice in programming assignments. J. Comput. Sci. Coll. **26**, 110–116 (2011)
9. Battestilli, L., Korkes, S.: Writing effective autograded exercises using bloom's taxonomy. In: 2020 ASEE Virtual Annual Conference Content Access Proceedings. ASEE Conferences (2020). https://doi.org/10.18260/1-2-35711
10. Roberts, T.S.: The use of multiple choice tests for formative and summative assessment. In: Proceedings of the 8th Australasian Conference on Computing Education – vol. 52 (ACE'06), pp. 175–180. Australian Computer Society, Australia (2006)
11. Sanders, K., et al,: The Canterbury Questionbank: building a repository of multiple-choice CS1 and CS2 questions. In: Proceedings of the ITiCSE working group reports conference on Innovation and technology in computer science education-working group reports – ITiCSE – WGR'13, pp. 33–52. ACM Press, New York, New York, USA (2013). https://doi.org/10.1145/2543882.2543885

12. Bower, M.: A taxonomy of task types in computing. In: Proceedings of the 13th Annual Conference on Innovation and Technology in Computer Science Education – ITiCSE'08, pp. 281–285. ACM Press, New York, New York, USA (2008). https://doi.org/10.1145/138 4271.1384346

13. Ruf, A., Berges, M., Hubwieser, P.: Classification of Programming Tasks According to Required Skills and Knowledge Representation. In: Brodnik, A. and Vahrenhold, J. (eds.) Informatics in Schools. Curricula, Competences, and Competitions – ISSEP 2015, Lecture Notes in Computer Science. pp. 57–68. Springer, Cham (2015). https://doi.org/10.1007/978-3-319-25396-1_6

14. Ruf, A., Berges, M., Hubwieser, P.: Types of assignments for novice programmers. In: Proceedings of the 8th Workshop in Primary and Secondary Computing Education on – WiPSE'13, pp. 43–44. ACM Press, New York, New York, USA (2013). https://doi.org/10.1145/2532748.2532777

15. Grant, M.J., Booth, A.: A typology of reviews: an analysis of 14 review types and associated methodologies. Health Info. Libr. J. **26**, 91–108 (2009). https://doi.org/10.1111/j.1471-1842.2009.00848.x

16. Clark, D.: Testing programming skills with multiple choice questions. Inform. Educ. **3**, 161–174 (2004)

17. Gomes, A., Correia, F.B.: Bloom's taxonomy based approach to learn basic programming loops. In: 2018 IEEE Frontiers in Education Conference (FIE), pp. 1–5. IEEE (2018). https://doi.org/10.1109/FIE.2018.8658947

18. Sheard, J., Simon, Dermoudy, J., D'Souza, D., Hu, M., Parsons, D.: Benchmarking a set of exam questions for introductory programming. In: Proceedings of the Sixteenth Australasian Computing Education Conference – vol. 148, pp. 113–121 (2014)

19. Arunoprayoch, N., Lai, C.-H., Tho, P.-D., Liang, J.-S., Yang, J.-C.: Effects of question types on engagement and performance of programming learning for non-computer science majors. In: 2018 7th International Congress on Advanced Applied Informatics (IIAI-AAI), pp. 306–311. IEEE (2018). https://doi.org/10.1109/IIAI-AAI.2018.00065

20. Sheard, J., Carbone, A., Lister, R., Simon, B., Thompson, E., Whalley, J.L.: Going SOLO to assess novice programmers. In: Proceedings of the 13th Annual Conference on Innovation and Technology in Computer Science Education, pp. 209–213. ACM, New York, NY, USA (2008). https://doi.org/10.1145/1384271.1384328

21. Chatzopoulou, D.I., Economides, A.A.: Adaptive assessment of student's knowledge in programming courses. J. Comput. Assist. Learn. **26**, 258–269 (2010). https://doi.org/10.1111/j.1365-2729.2010.00363.x

22. Malik, S.I., Tawafak, R.M., Shakir, M.: Aligning and assessing teaching approaches with SOLO taxonomy in a computer programming course. Int. J. Inf. Commun. Technol. Educ. **17**, 1–15 (2021). https://doi.org/10.4018/IJICTE.20211001.oa5

23. Malik, S.I.: Assessing the teaching and learning process of an introductory programming course with bloom's taxonomy and assurance of learning (AOL). Int. J. Inf. Commun. Technol. Educ. **15**, 130–145 (2019). https://doi.org/10.4018/IJICTE.2019040108

24. Omar, N., et al.: Automated analysis of exam questions according to bloom's taxonomy. Procedia Soc. Behav. Sci. **59**, 297–303 (2012). https://doi.org/10.1016/j.sbspro.2012.09.278

25. Gomes, A., Correia, F.B., Abreu, P.H.: Types of assessing student-programming knowledge. In: 2016 IEEE Frontiers in Education Conference (FIE), pp. 1–8. IEEE (2016). https://doi.org/10.1109/FIE.2016.7757726

26. Sindre, G.: Code writing vs code completion puzzles: analyzing questions in an E-exam. In: 2020 IEEE Frontiers in Education Conference (FIE), pp. 1–9. IEEE (2020). https://doi.org/10.1109/FIE44824.2020.9273919

27. Abreu, P.H., Silva, D.C., Gomes, A.: Multiple-choice questions in programming courses: can we use them and are students motivated by them? ACM Trans. Comput. Educ. **19**, 1–16 (2019). https://doi.org/10.1145/3243137
28. Izu, C., Weerasinghe, A., Pope, C.: A study of code design skills in novice programmers using the SOLO taxonomy. In: Proceedings of the 2016 ACM Conference on International Computing Education Research, pp. 251–259. ACM, New York, NY, USA (2016). https://doi.org/10.1145/2960310.2960324
29. Masapanta-Carrión, S., Velázquez-Iturbide, J.Á.: Evaluating instructors' classification of programming exercises using the revised bloom's taxonomy. In: Proceedings of the 2019 ACM Conference on Innovation and Technology in Computer Science Education, pp. 541–547. ACM, New York, NY, USA (2019). https://doi.org/10.1145/3304221.3319748
30. Harley, Z., Harley, E.: E-learning and e-assessment for a computer programming course. In: 3rd International Conference on Education and New Learning Technologies, pp. 2074–2080. Barcelona, Spain (2011)
31. Dorodchi, M., Dehbozorgi, N., Frevert, T.K.: "I wish I could rank my exam's challenge level!": an algorithm of Bloom's taxonomy in teaching CS1. In: 2017 IEEE Frontiers in Education Conference (FIE), pp. 1–5. IEEE (2017). https://doi.org/10.1109/FIE.2017.8190523
32. Lajis, A., Nasir, H.M., Aziz, N.A.: Proposed assessment framework based on bloom taxonomy cognitive competency. In: Proceedings of the 2018 7th International Conference on Software and Computer Applications, pp. 97–101. ACM, New York, NY, USA (2018). https://doi.org/10.1145/3185089.3185149
33. Garcia, R., Falkner, K., Vivian, R.: Instructional framework for CS1 question activities. In: Proceedings of the 2019 ACM Conference on Innovation and Technology in Computer Science Education, pp. 189–195. ACM, New York, NY, USA (2019). https://doi.org/10.1145/3304221.3319732
34. Denny, P., Luxton-Reilly, A., Hamer, J., Purchase, H.: Coverage of course topics in a student generated MCQ repository. In: Proceedings of the 14th annual ACM SIGCSE conference on Innovation and technology in computer science education, pp. 11–15. ACM, New York, NY, USA (2009). https://doi.org/10.1145/1562877.1562888
35. Simon, et al.: Introductory programming: examining the exams. In: Proceedings of the Fourteenth Australasian Computing Education Conference (ACE '12) – vol. 123. pp. 61–70. Australian Computer Society, Inc., AUS (2012)

Beyond the Classroom

An Alumni Satisfaction Model for Computing Departments

Andre P. Calitz[1]([✉]) [iD], Margaret Cullen[2] [iD], Arthur Glaum[2] [iD], and Jean Greyling[1] [iD]

[1] Department of Computing Sciences, Nelson Mandela University, Port Elizabeth, South Africa
{Andre.Calitz,Jean.Greyling}@Mandela.ac.za
[2] Business School, Nelson Mandela University, Port Elizabeth, South Africa
Margaret.Culen@Mandela.ac.za

Abstract. The perception of Alumni about the extent of learning and the usefulness of the knowledge gained is a key measure for education institutions to assess their success. Alumni are important stakeholders, specifically for Higher Education Institutions, as they work in industry and can provide valuable feedback on education and service offerings, including course content. The importance of Alumni satisfaction in relation to the success of universities makes it imperative that the factors driving satisfaction be determined. The purpose of this paper is to measure Alumni satisfaction in a department offering Computer Science and Information Systems qualifications and identify areas of improvement.

A hypothesised model of the factors influencing Alumni satisfaction was proposed and used to guide the quantitative research study. The research was exploratory and eight factors affecting Alumni satisfaction were investigated. A questionnaire was distributed to department's Alumni who work in the IT industry. The results indicate that the Alumni agreed that the academic staff maintained high academic standards, the degree programmes and modern technologies prepared them for the world of work. One factor that should be improved upon is departmental communications with the Alumni network. Technology and education are rapidly changing and departments must regularly determine Alumni satisfaction.

Keywords: Alumni satisfaction · Service guarantee · Stakeholder theory · Perceived value

1 Introduction

Alumni are considered the most important assets of a university or any academic institution [1–3]. Alumni are important assets for any academic institution because the institution is represented in the real working world by their Alumni. The achievements of Alumni directly reflect on the academic institution or academic department and any improvements to the quality of the education automatically improves the perceived value of the graduate's qualification [4].

The success of courses offered by a university and the effectiveness of the lecturers and the support staff have been measured by student evaluations [5–7]. The feedback generated is typically used to identify lecturers who need additional training, courses

H. E. Van Rensburg et al. (Eds.): SACLA 2023, CCIS 1862, pp. 137–151, 2024.
https://doi.org/10.1007/978-3-031-48536-7_10

that need to be restructured or the need for financial rewards for those who excel. The principal objective of educating students is the knowledge and skills they can use and apply after graduating, regardless of the field of study. The Alumni perception of the extent of learning and the usefulness of the knowledge is a key measure for universities [6, 7].

Alumni satisfaction, seen as a representation of higher education institution quality, is determined by the qualifications and professionalism of lecturers and the quality of facilities and infrastructure services [8]. Alumni satisfaction can further be defined as the Alumni's perception of contentment with the academic institution, their experiences whilst at the institution and the extent to which they feel their learning expectations were met while studying at the academic institution [9]. The factors that influence Alumni satisfaction include the lecturer's professionalism, the relevance of the curriculum, the quality and availability of facilities and infrastructure [8]. The study by Petratos and Calitz [1] identified the following factors relating to Alumni satisfaction at a tertiary institution offering IT related programmes, Customer Satisfaction, Course Contents, Modern Technologies, Academic Staff, Administrative and Technical Staff, Social Environment, Perceived Value and Alumni Network.

McCollough and Gremler [10] define a service guarantee as a formal promise made to customers about the service they will receive. Similarly, McColl and Mattsson [11] define a service guarantee as a written promise made by the company through advertising or company literature that will provide compensation if promises are broken. In order to ensure customer satisfaction in a service offering, the quality of service performance needs to be guaranteed [11, 12]. A satisfied customer can be a potential carrier of positive word of mouth, where an optimistic social image could be a benefit that universities realise [13]. Service guarantee was included as an additional independent factor relating to Alumni satisfaction in this study.

Stakeholder theory suggests that the purpose of an organisation is to create as much value as possible for stakeholders and not only for the shareholders. Alumni are important stakeholders for universities as they provide valuable financial, intellectual and human resources and enhance the universities image and reputation [14]. An organisation should try to meet the needs of everyone who has a stake in both the actions and outcomes of the organisation [15]. An organisational competitive advantage is gained by involving the stakeholders, as strategic resources, in corporate decisions [16].

The research objective, research problem and the Alumni satisfaction questionnaire are discussed in Sect. 2. Literature on the factors that influence Alumni satisfaction are discussed in Sect. 3. The Alumni satisfaction survey results are presented in Sect. 4. Conclusions, recommendations of this study and future work are discussed in Sect. 5.

2   The Research Objective, Problem and Research Design

The objective of this paper was to explore and report on the factors influencing Alumni satisfaction in a Computer Science (CS) and Information Systems (IS) Department at a comprehensive university offering under-graduate and post-graduate degree programmes in South Africa. The intention of the research was to study the level of Alumni satisfaction with the educational programme, as well as to analyse the quality of professional programmes and the educational process in the department.

The research problem investigated in this study was that the department has not determined the satisfaction levels of Alumni working in industry. An in-depth literature review of studies related to the factors that influence the perceptions of Alumni was undertaken. A hypothesised model of Alumni satisfaction was derived from the literature reviewed (Fig. 1). The model was then used to develop a questionnaire, which was distributed to the Alumni of the Department of Computing Sciences.

2.1 Participants

The sampling method used in this study was purposeful sampling as the respondents consisted of Alumni whose contact information was obtained from the database of the Department of Computing Sciences. The data were collected from the sample by means of an online questionnaire. An email containing a Universal Resource Link (URL) to the questionnaire was sent to the email addresses of the Alumni of the Department of Computing Sciences and there were more than 600 potential respondents working in the IT industry. The URL was also posted on the Facebook page of the department. The potential respondents were reminded to respond three times after which a total of 100 fully completed responses were received.

2.2 Measuring Instrument

The Alumni satisfaction questionnaire consisted of seven sections and was divided into fourteen sub-sections with a total of 50 questions and items. Section A of the questionnaire captured the biographical information of the respondents. Section A also captured other information designed to gauge the respondents' level of education. Information pertaining to the respondents' organisation, such as the size of the company where they work, if they were self-employed or not and the total number of years worked in the IT industry. This section contained a total of fourteen questions.

Section B to Section G captured the respondents' perception of Service Guarantees, Customer Satisfaction, Course Contents, Modern Technologies, Academic Staff, Administrative and Technical Staff, Social Environment, Perceived Value and Alumni Network all relating to the Alumni Satisfaction. These items were rated using a five-point Likert scale where 1 = Strongly Disagree and 5 = Strongly Agree. This section measured a total of 10 factors. Each factor was measured using between four and eight items.

Examples of items for factors rated on the five-point Likert scale are as follows:

- Service Perception – "The service quality at Department of Computer Sciences met my expectations".
- Technology – "Technology utilised at the Department of Computer Sciences was kept up to date".
- Academic staff – "The academic staff were adequately qualified".

2.3 Data Analysis Methods

The hypotheses developed in this quantitative study were tested statistically. The university research statistician used the software package STATISTICA. The statistical analyses included descriptive statistics and inferential statistics, specifically Pearson's correlation analysis.

2.4 Pilot Study

The questionnaire, developed from literature was pilot tested by eight graduates working in the Port Elizabeth area. The university research statistician also provided feedback and minor changes were made to the questionnaire used in this research study.

3 Literature Review

Analysing the stakeholders of the university can provide insight into the quality of programmes offered by academic departments, the perceived value of academic offerings and the relevance of the programmes on offer. Investigating the satisfaction of a department's Alumni can provide valuable feedback on the Alumni's experience and perceived value of the degree and diploma programmes they completed [1]. Departments can use this feedback to assess how relevant the course on offer is and if the experience of the Alumni was positive. Improvements, interventions and enhancements can be made by departmental management to improve the academic offerings and programmes. Existing factors affecting Alumni satisfaction, including the new factor, Service Guarantee that were used in this study are presented below.

Service Guarantees
An implicit assumption in the services guarantee literature is that offering a service guarantee will increase customer satisfaction [12, 17, 18]. Service guarantees give rise to increased customer satisfaction in several ways [5, 11]. Organisations that are more focused on customers by providing guarantees can realise many benefits. These benefits include clear standards that are set for the organisation with guarantees [12]. Feedback is generated from a guarantee because dissatisfaction from customers is expressed when guarantees are not met [3, 19]. Failure is measured against a guarantee creating greater understanding [10, 12]. There are marketing benefits derived from guarantees as they reduce purchasing risk and enhance the loyalty of existing customers [10].

The perception of guarantees is that higher guarantees are signals of higher quality and lower guarantees are signals of lower quality [11, 12]. Based on customer perception, implementing a service guarantee can either be a positive or negative signal. However, Hogreve and Gremler [20] suggest that providing the service guarantee will result in internal changes, which will improve quality and ultimately improve customer perception and the value of the guarantee.

Unlike businesses, universities rarely guarantee the services they offer to students. One of the reasons identified is because services, especially education, are considered intangible [10, 20]. The intangible quality of education is seen as riskier and failure is more common with an intangible product than with tangible product failure. Firstly, if failure is more likely, then service guarantees are also perceived to be more expensive. Secondly, the design and drafting of a service guarantee is more challenging. Lastly, due to co-production it is difficult to separate the roles and responsibilities of the consumer and service provider [10].

Customer Satisfaction

Customer value is a complex concept as it is often interpreted with various meanings depending on the point of view adopted. A definition of customer value is the perceived value that the customer gains when purchasing a product or receiving a service [17, 21, 22]. The customer perceives value in the product or service when the benefits exceed the costs. Alumni are customers and academic institutions are trying to improve customer satisfaction by improving the quality of the service provided to students [23].

The conventional method of listening to the customers' needs has been achieved through measuring customer satisfaction. Having a better understanding of what customers value will help organisations to achieve their organisational purpose and goals. This has resulted in an extended view of customer-perceived value, which has drawn the attention of researchers [17, 21, 22]. Based on this extended view the focus on customer value then changes to a customer-oriented concept. An organisation that is customer-oriented must consider the perception of the customer when it defines the value proposition of the organisation.

Course Content

Potential students are influenced by various factors when selecting a course or registering for a degree programme. Factors that have an influence are interactions with staff during the application process and at open days, university location, clarity and quality of printed materials and the course content [24]. The course content is of major importance to attract new students for a qualification. Alumni, from industry, are an excellent source of feedback as their experience in industry can be related directly back to course content and assessments [25]. It is essential to evaluate the curriculum according to customer satisfaction by students and Alumni, to revise the curriculum [8]. The attributes of the academic programme and course content were the most significant factors in Wiranto and Slamento's [8] study of student satisfaction.

Modern Technologies

The quality of technologies used in a tertiary institution is important to ensure student satisfaction. Universities must strategically use modern technologies to support teaching and learning in IT. There are various reasons why universities would increase IT for curriculum delivery, but the quality of teaching is the central concern of lecturers. There must be a specific link between using relevant technology during course work and the technology that will be used in the work environment [1]. More importantly, the choice of IT used and the way it is designed into study activities will create a positive attitude and efficient, effective performance towards IT. Newer students may take existing technology for granted and the attitudes towards the usefulness and ease of use will play a strong role on willingness to develop new skills and technology usage [26, 27]. Technology and infrastructure as part of the departmental facilities can improve the rate of student and Alumni satisfaction [8].

Academic Staff

A study conducted by Xulu-Gama et al. [27] indicated that students reacted positively in an environment where academic staff (lecturers) gave them attention, cared about them and were friendly. The success of a course offered by a university and the effectiveness of the instructors have commonly been measured by means of student evaluations [5–7]. The feedback generated is typically used to identify lecturers who need additional training, courses that need to be restructured or the need for financial rewards for those who excel. The principal objective of educating students is to inculcate what they can use and apply after graduating, regardless of the field of study. The Alumni perception of the extent of learning and the usefulness of the knowledge is a key measure for universities [6, 7]. Academics are professionals and need to have several competencies, including having the relevant qualifications; understanding students; teaching, learning and education; personal competence and professional development [8].

Administrative and Technical Staff

A wide variety of duties in a university is taken care of by the administrative and technical staff in departments. The quality of administrative and technical staff is often downplayed at HEIs. University staff are required to be dedicated, efficient and need to have the ability to work both in a team and individually. It is difficult to categorise and grade the work as it is both varied and demanding [8]. In terms of the stakeholder model, administrative staff are viewed as secondary stakeholders. The primary stakeholders are those without whom the organisation cannot exist or those who have an official or contractual relationship with the organisation. Administrative staff are key in the interface with students and contribute to a positive or negative experience at the university [15]. CS and IS students work in university computer laboratories during their studies and require assistance from technical personnel. The quality and experience of the technical staff and the service quality they provide to students is of vital importance [1].

Social Environment

According to Pedro, Pereira and Carrasqueira [28], the social environment has been identified as an important factor influencing Alumni satisfaction. The social environment refers to the various relationships that Alumni may have fostered throughout their academic life, as well as the evaluation of the extramural activities that may influence these relationships. Pedro et al. [28] found that the relationship between the university and students influences the longevity of the relationship in the future and the commitment of Alumni.

Perceived Value

In order to achieve a competitive advantage and predict customer behaviour, much attention has been given by both academics and organisations to the value perceived by the customer [23]. The perceived value of a product or a service will vary from one customer to the next. Tailoring a product or service to a wide audience will result in some customers valuing a certain aspect more when compared to other customers. The value of a product or service will also change over time as customers' needs and wants change. The perceived value by the customer of the product or service on offer is difficult to measure [17, 23].

The perceived value of an academic qualification has a direct impact on how the institution is portrayed by the student or Alumni to others in the industry. This perceived value thus plays a vital role in the competitive advantage institutions with similar offerings would have [3, 19]. The achievements of Alumni directly reflect on the department and the university and any improvements to the quality of the education at the university automatically improve the perceived value of the graduate's qualification [4]. Alumni are a university's best ambassadors and should be kept informed and involved using the Alumni network.

Alumni Network
Historically, Alumni networks were created from regional groups brought together for fundraising purposes [19]. Over time these networks developed, both in their importance and benefits and the department and university gained from these networks in terms of the development of the university. Alumni groups are constantly evolving and have been in existence for decades. Changes have been accelerated in recent years with the development of the Internet and social networking that facilitates global communication. The COVID-19 pandemic period required different communication tools to be used to interact with students and Alumni [30]. This is another reason why the Alumni networks are vital for universities to enhance their growth and development. The Internet and social media, specifically Facebook and LinkedIn have assisted Alumni networks to remain in contact with one-another and build and maintain friendships and professional relationships. Departments have further used the Alumni network to maintain relationships, obtain funding and to provide a platform for Alumni and student interaction [1].

3.1 Hypotheses for the Alumni Satisfaction Model

A hypothesised model, based on the literature review, was developed for this research study (Fig. 1). The hypothesised model was used to investigate the relationships between the dependent factor, namely Alumni satisfaction and the independent factors: Service guarantees, Customer satisfaction, Course contents, Modern technologies, Academic staff, Administrative staff, Social environment, Perceived value and Alumni network. The hypotheses developed in this research study were formulated to be proven true or false by means of statistical analysis through empirical evaluation and to verify the proposed relationships indicated in the hypothesised model.

The following hypotheses were formulated to test the relationship between the Independent factors and the Dependent factor:

H_1: Service Guarantees are significantly related to Alumni Satisfaction.
H_2: Customer Satisfaction is significantly related to Alumni Satisfaction.
H_3: Course Content is significantly related to Alumni Satisfaction.
H_4: Modern Technologies are significantly related to Alumni Satisfaction.
H_5: Academic Staff are significantly related to Alumni Satisfaction.
H_6: Administrative Staff are significantly related to Alumni Satisfaction.
H_7: Social Environment is significantly related to Alumni Satisfaction.
H_8: Perceived Value is significantly related to Alumni Satisfaction.
H_9: Alumni Network is significantly related to Alumni Satisfaction.

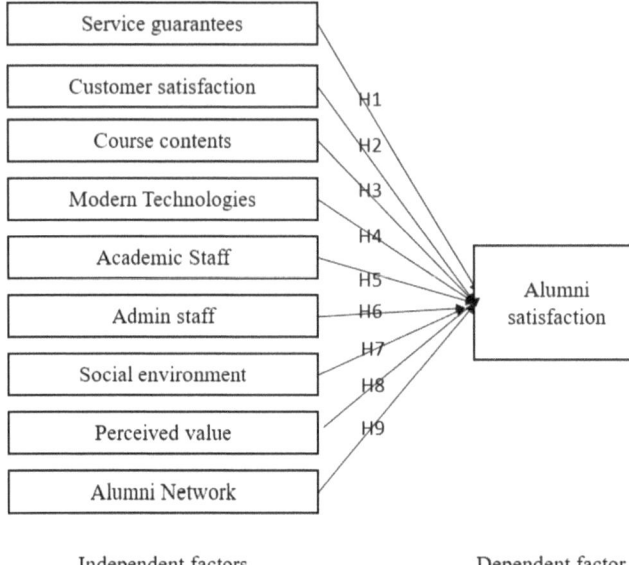

Independent factors   Dependent factor

Fig. 1. Hypothesised Alumni Satisfaction Model

4 Results and Discussion

Section A of the questionnaire enabled descriptive statistics to be measured in the form of biographical information, including Gender, Age, Education Level and Geographical place of residence (Table 1). Section A also captured information related to employment, including size of the company they work for, if they are self-employed or not and the years in the IT industry (Table 2).

The result indicated that 82% of the respondents were male and 18% were female (Table 1). Of the 100 respondents, 19% of the respondents are between the ages of 20–24, 43% between 25–34, 28% between 35–44, seven between 45–54, and three were 55 years of age or older. The results show that 47% of the respondents reside in the Eastern Cape, 12% reside in Gauteng, 21% reside in Western Cape and 20% reside in a country outside South Africa. No responses were received for respondents residing in the Free State, KwaZulu-Natal, Limpopo, Mpumalanga, Northern Cape and North West.

The respondents were asked to indicate their undergraduate qualifications. The results show that 34% of the respondents qualified in BSc CS, 14% in BSc IS, 30% in BCom CS and IS, nine in BCom IS, six in BCom Rat and seven selected Other. The respondents were asked to indicate the year in which they completed their undergraduate degree. The results indicate that many of the respondents (63%) graduated between the years 2010 and 2019.

The results show that 44 of the respondents who completed a postgraduate degree, have an honours qualification, 14 a masters degrees and six a doctorate. The respondents were asked to indicate the year in which they completed their postgraduate degree. The results indicate that many of the respondents 66 graduated between the years of 2010 and 2019. The respondents were asked to provide any additional degrees they have

Table 1. Demographic profile of participants ($n = 100$)

Education	UG	Honours	Masters	PhD
	36%	44%	14%	6%
Age	*20–24 years*	*25–34 years*	*35–44 years*	*45+ years*
	19%	43%	28%	10%
Place of residence	*EC*	*Gauteng*	*WC*	*Outside SA*
	47%	12%	21%	20%
Qualification completed	*1974–1984*	*1985–1994*	*1995–2004*	*2005–2019*
	4%	7%	26%	63%
Gender	*Male*	*Female*	*Total*	
	82%	18%	100 (100%)	

completed, other than those awarded by the Department of Computing Sciences. The results show that fifteen respondents completed a second qualification. The most popular second degree was a Master of Business Administration (MBA).

Table 2 indicates that of the 100 respondents, 41 have fewer than 5 years of experience in IT, 20 between 5–9 years, 28 between 10–19 years and eleven 20 years or more experience in IT. Regarding the respondents' organisation where they are employed, there was a wide range of organisations, from multinational corporations to small businesses. The more popular organisations to work for were Amazon, Lightstone Consumer, NMU, Openbox Software, SYSPRO and S4 Integration. The results show that two of the respondents are employed in the Government sector, nine in Manufacturing, nine in Financial Services, two in Pharmaceutical, one in Agriculture, one in Mining, four in Education, one in Health Services, two in Retail, seven in Services, forty-five in Information Technology/Telecommunications, one in Electricity/Water Services and 16 indicated a sector other than those listed in the questionnaire. Of the 100 respondents, 19 work for companies with 1–9 employees, 43 with between 10–99 employees, 28 with between 100–999 employees, seven with between 1000–4999 employees and three with 5000+ employees. Eight respondents were self-employed in the IT industry.

4.1 Reliability

Cronbach's Alpha coefficient is a measure of internal consistency of the research instrument. The reliability of the questionnaire for the factors Customer Satisfaction, Academic Staff, Course Contents, Alumni Network and Alumni Satisfaction were *excellent* ($\alpha >= 0.9$). The Cronbach Alpha coefficients for the factors Perceived Value, Modern Technologies, Administrative Staff and Social Environment indicated a *good* reliability ($0.9 > \alpha >= 0.8$) [29]. A Cronbach's Alpha value of between 0.50 and 0.69 has been deemed acceptable for new and experimental research [29]. Service Guarantees was included as a new factor relating to Alumni Satisfaction is thus seen as *acceptable* ($\alpha = 0.69$) (Table 3).

Table 2. Employment information ($n = 100$)

Employment information		Percentage
Years in ICT industry	<5 years	41
	5–9 years	20
	10–19 years	28
	20+ years	11
Industry sector	Government	2
	Manufacturing	9
	Financial Services	9
	Pharmaceutical	2
	Agriculture	1
	Mining	1
	Education	4
	Management Consulting	0
	Health Services	1
	Retail	2
	Services	7
	IT/Telecommunications	45
	Electricity/Water Services	1
	Other	16
Number of employees at company	1–9 employees	9
	10–99 employees	23
	100–999 employees	30
	1000–4999 employees	16
	5000+ employees	22
Self-employed	No	92
	Yes	8

4.2 Correlations and Hypotheses Testing

The correlations proved to be both statistically and practically significant at a 0.05 confidence level when rcrit is bigger or equal to 0.300 for all correlations. There are very high positive correlations between Customer satisfaction, Course contents, Modern technologies, Academic staff, Administrative staff and perceived value with Alumni satisfaction (Table 4). These strong positive correlations are aligned with the literature investigated. There are also medium positive correlations between Social environment and Alumni network with Alumni satisfaction. These medium positive correlations are

Table 3. Cronbach Alpha ($n = 100$)

Factor	Cronbach Alpha	Reliability
Customer Satisfaction	0.97	Excellent
Academic Staff	0.94	Excellent
Course Contents	0.92	Excellent
Perceived Value	0.84	Good
Modern Technologies	0.88	Good
Administrative Staff	0.87	Good
Social Environment	0.87	Good
Alumni Network	0.91	Excellent
Service Guarantees	0.69	Acceptable
Alumni Satisfaction	0.91	Excellent

also aligned with the literature, but changes in these factors have a smaller influence on Alumni satisfaction.

A significant finding is the strong positive relationship ($r = 0.881$) between Customer satisfaction and Alumni satisfaction. It can be deduced that a student, as a satisfied customer, becomes a satisfied alumnus. This relationship is expected to have a strong correlation as both are related to satisfaction. The factor concerning Academic staff has a strong positive relationship ($r = 0.880$) with Alumni satisfaction. This finding supports the reviewed literature where a positive relationship between the student perceptions of learning and the ranking of the lecturer was identified by Wiranto and Slameto [8].

The hypothesised Alumni model was constructed based on the literature (Fig. 1). The hypothesised Alumni model was used to establish relationships between the dependent factor, namely Alumni satisfaction, and the independent factors: Service guarantees, Customer satisfaction, Course contents, Modern technologies, Academic staff, Administrative staff, Social environment, Perceived value and Alumni network using Pearson correlations. Eight out of the nine hypotheses developed in this study were accepted by means of statistical analysis through empirical evaluation (Table 4). The model therefore needs to be adjusted by only removing H_1 as an independent factor of Alumni Satisfaction.

4.3 Discussion

The available course content at the Department of Computing Sciences has a strong positive relationship ($r = 0.874$) with Alumni satisfaction. This finding supports the reviewed literature where students are influenced by various factors when selecting a course [24]. The perceived value obtainable at the Department of Computing Sciences has a strong positive relationship ($r = 0.862$) with Alumni satisfaction. This finding supports the reviewed literature where the achievements of Alumni directly reflect on the university and any improvements to the quality of the education at the university automatically improves the perceived value of the graduate's degree [4].

Table 4. Correlations and hypotheses testing results

Hypotheses	Relationship	Pearson Correlations	Correlation Strength	Remarks
H_1	*Service Guarantees → Alumni Satisfaction*	0.204	Low positive correlation	Rejected
H_2	*Customer Satisfaction → Alumni Satisfaction*	0.881	High positive	Accepted
H_3	*Course Content → Alumni Satisfaction*	0.874	High positive	Accepted
H_4	*Modern Technologies → Alumni Satisfaction*	0.828	High positive	Accepted
H_5	*Academic Staff → Alumni Satisfaction*	0.880	High positive	Accepted
H_6	*Administrative Staff → Alumni Satisfaction*	0.812	High positive	Accepted
H_7	*Social Environment → Alumni Satisfaction*	0.773	Medium positive	Accepted
H_8	*Perceived Value → Alumni Satisfaction*	0.862	High positive	Accepted

The modern technologies on offer at the Department of Computing Sciences have a strong positive relationship ($r = 0.828$) with Alumni satisfaction. This finding supports the literature stating that universities strategically use technology to support learning in IT [1]. The choice of IT used and the way it is designed into study activities will create a positive attitude and efficient effective performance in IT. Newer students may take existing technology for granted and the attitudes towards the usefulness and ease of use will play a strong role on willingness to develop new skills and technology usage.

The factor involving administrative and technical staff at the Department of Computing Sciences has a strong positive relationship ($r = 0.812$) with Alumni satisfaction. This finding supports the reviewed literature where administrative and technical staff are considered an interface with students and contribute to a positive or negative experience at the university [15]. The factor concerning the social environment at the Department of Computing Sciences has a medium positive relationship ($r = 0.773$) with Alumni satisfaction. This positive correlation supports a well-designed social environment, not only at the university, but also with the surrounding facilities.

The alumni network at the department of Computing Sciences has a medium positive relationship ($r = 0.701$) with Alumni satisfaction. This positive correlation is supported by the literature investigated [3, 19]. Service guarantees have a low positive correlation ($r = 0.204$) and any change in this factor is unlikely to influence Alumni satisfaction. The low positive correlation is in contrast with the literature investigated. Service guarantees are not well known to the respondents and they felt that it would have no influence on Alumni satisfaction. From the literature, service guarantees offered by a university can still generate feedback if there is dissatisfaction and failure [10, 23]. Lecturers can learn from students and adapt the course to service future students, who invoke guarantees, better. The increasing focus on student evaluations and teaching quality highlights the need to understand student dissatisfaction, which can be better understood by offering a service guarantee.

5   Conclusions, Limitations and Future Research

Research has indicated that the main determinant of Alumni satisfaction is lecturer professionalism [8]. Tangible factors, such as physical facilities, equipment and communication facilities also have a significant effect on Alumni satisfaction [23]. This study aimed to investigate and report on the factors influencing Alumni satisfaction at a HEI offering CS and IS degree programmes. A literature review was conducted to develop an understanding of Alumni satisfaction and a conceptual model of Alumni satisfaction was developed and hypothesised. The relationships between the dependent factor of Alumni satisfaction and the independent factors, namely Service Guarantees, Customer Satisfaction, Course Contents, Modern Technologies, Academic Staff, Administrative and Technical Staff, Social Environment, Perceived Value and Alumni Network all relating to the Alumni Satisfaction were tested using Pearson correlations.

All the independent factors had high to medium correlations with the dependent factor Alumni Satisfaction. The only factor that had a low positive correlation ($r = 0.204$) was Service Guarantees. An important finding was the strong positive relationship between customer satisfaction and Alumni satisfaction. The Department of Computing Sciences should allocate resources and prioritise customer satisfaction as any perceived change would have a significant effect on Alumni satisfaction.

The communication methods used by the Department of Computing Sciences were identified as a mismatch. The communication channels used most frequently by the respondents are email, social media, SMS and mobile phones. A recommendation to involve the alumni and improve the Alumni network was that these channels should be used more frequently to contact Alumni. The Alumni portal was not used often and currently was not an effective way to communicate with the Alumni. The Department of Computing Sciences needs to create greater awareness of this communication method or use it as a secondary communication channel.

The factors that affect Alumni Satisfaction has changed the past years, due to the COVID-19 pandemic and working online [30]. None of the respondents who participated in this study, studied CS and IS during the pandemic period. Future research must include responses from Alumni who studied during the pandemic period. The on-line, working-from-home learning environment provided additional challenges, which requires different teaching and support services. A larger sample size will provide advanced statistical analysis where the model can be further evaluated using Exploratory Factor Analysis. Further in-depth research could also be conducted to understand why there are differences between the perception of respondents only having a graduate degree and respondents having a postgraduate qualification. The factor Employability must also be included in the Alumni satisfaction model and further explored. Teaching methods, technology and CS and IS education are rapidly changing and computing departments must determine Alumni satisfaction on a three year cycle.

Acknowledgement. The paper is based on post-graduate research conducted by Glaum [31] at the Nelson Mandela University Business School.

References

1. Petratos, S., Calitz, A.P.: Evaluating alumni satisfaction in the school of ICT. In: SACLA 2019, the 48th Annual Conference of the Southern African Computer Lecturers' Association Conference, Alpine Heath Resort, Drakensberg, S.A., 15–17 July (2019)
2. Masserini, L., Bini, M., Pratesi, M.: Do quality of services and institutional image impact students' satisfaction and loyalty in Higher Education? Soc. Ind. Res. **146**(1–2), 91–115 (2018)
3. Rattanamethawong, V., Sinthupinyo, S., Chandrachai, E.A.: An innovation system that can quickly responses to the needs of students and alumni. Procedia – Soc. Behav. Sci. **182**, 645–652 (2015)
4. Egizii, R.: Self-directed learning, andragogy and the role of alumni as members of professional learning communities in the post-secondary environment. Procedia – Soc. Behav. Sci. **174**, 1740–1749 (2015)
5. Guevara, C., Stewart, S.: Do student evaluations match alumni expectations? Manag. Financ. **37**(7), 610–623 (2011)
6. Khatimin, N., Wahab, D.A., Mohamed, A.: Postgraduate alumni survey of the faculty of engineering and built environment. Procedia – Soc. Behav. Sci. **18**, 110–117 (2011). https://doi.org/10.1016/j.sbspro.2011.05.016
7. McDearmon, J.: Hail to thee, our alma mater: alumni role identity and the relationship to institutional support behaviors. Res. High. Educ. **54**(3), 283–302 (2013)
8. Wiranto, R., Slameto, S.: Alumni satisfaction in terms of classroom infrastructure, lecturer professionalism, and curriculum. Heliyon **7**(6), e06679 (2021). https://doi.org/10.1016/j.heliyon.2021.e06679
9. Naeem, I., Aparicio-Ting, F.E., Dyjur, P.: Student stress and academic satisfaction: a mixed methods exploratory study. Int. J. Innov. Bus. Strategies (IJIBS) **6**(1), 388–395 (2020)
10. McCollough, M.A., Gremler, D.D.: Guaranteeing Student Satisfaction: An Exercise in Treating Students as Customers. J. of Mark. Educ. **21**(2), 118–130 (1999)
11. McColl, R., Mattsson, J.: Common mistakes in designing and implementing service guarantees. J. Serv. Mark. **25**(6), 451–461 (2011)
12. Tucci, L.A., Talaga, J.: Service guarantees and consumers' evaluation of services. J. Serv. Mark. **11**(1), 10–18 (1997)
13. Lukić, V.R., Lukić, N.: Assessment of student satisfaction model: evidence of Western Balkans. Total Quality Man. **31**, 1506–1518 (2018)
14. Tulankar, S., Grampurohit, B.P.: Role of alumni as stakeholders in enhancing quality education. In: Sustaining Quality: NAAC New Guidelines 2017 Conference, Bhandup, Mumbai (2020)
15. Caballero, G., Vázquez, X.H., Quintás, M.A.: Improving employability through stakeholders in European higher education: the case of Spain. Long Range Plan. **48**(6), 398–411 (2015)
16. Plaza-úbeda, J.A., Burgos-Jiménez, J., Carmona-Moreno, E.: Stakeholder integration: measuring and knowledge, interaction behavior dimensions adaptational. J. Bus. Ethics **93**(3), 419–442 (2014)
17. Dovaliene, A., Masiulyte, A., Piligrimiene, Z.: The relations between customer engagement, perceived value and satisfaction: the case of mobile applications. Procedia – Soc. Beh. Sc. **213**, 659–664 (2015)

18. Schofield, P., Fallon, P.: Assessing the viability of university alumni as a repeat visitor market. Tourism Man. **33**(6), 1373–1384 (2012)

19. Chi, H., Jones, E.L., Grandham, L.P.: Enhancing mentoring between alumni and students via smart alumni system. Procedia Comp. Sc. **9**, 1390–1399 (2012)

20. Hogreve, J., Gremler, D.D.: Twenty years of service guarantee research: a synthesis. J. Serv. Res. **11**(4), 322–343 (2009)

21. Song, H., Cadeaux, J., Yu, K.: The effects of service supply on perceived value proposition under different levels of customer involvement. Ind. Mark. Man. **54**, 116–128 (2015). https://doi.org/10.1016/j.indmarman.2015.12.003

22. Wouters, M., Kirchberger, M.A.: Customer value propositions as interorganizational management accounting to support customer collaboration. Ind. Mark. Man. **46**, 54–67 (2015)

23. Sudjoko, S., Masrum, K.: Alumni satisfaction in educational institutions: does the quality service effect? J. High. Educ. Theory and Prac. **22**(16), 208–216 (2022)

24. Brown, C., Varley, P., Pal, J.: University course selection and services marketing. Mark. Intell. Plan. **27**(3), 310–325 (2009)

25. Steele, A., Cleland, S.: Staying linkedIn with ICT graduates and industry. In: ITX Conference, New Zealand, 8–10 (2014)

26. Calitz, A.P., Greyling, J.H., Cullen, M.D.M.: South African industry ICT graduate skills requirements. South Afr. Comput. Lecturers' Assoc. (SACLA) **1**, 25–26 (2014)

27. Xulu-Gama, N., Nharib, S.R., Alcock, A., Cavanagh, M.: A student-centred approach: a qualitative exploration of how students experience access and success in a South African University of Technology. J. High. Educ. Res. and Dev. **37**(6), 1302–1314 (2018)

28. Pedro, I.M., Pereira, L.N., Carrasqueira, H.B.: Determinants for the commitment relationship maintenance between the alumni and the alma mater. J. Mark. High. Educ. **28**, 128–152 (2017)

29. Collis, J., Hussey, R.: Business Research: A Practical Guide for Undergraduate and Postgraduate Students, 3rd edn. Palgrave Macmillan, UK (2009)

30. Bugaj, J.M., Rybkowski, R.: Managing alumni loyalty. Poland from an international perspective. Polish J. Man. Stud. **26**(2), 75–91 (2022). https://doi.org/10.17512/pjms.2022.26.2.05

31. Glaum, A.: Alumni Perception of the NMU Computing Sciences Department MBA Treatise. Nelson Mandela University Business School, Port Elizabeth, South Africa (2018)

Motivations and Experiences of Recognition of Prior Learning Candidates in Information Systems Programmes

Thelma. M. T. Chitsa[(✉)] [iD] and Gwamaka Mwalemba[iD]

Department of Information Systems, Faculty of Commerce, University of Cape Town, Rondebosch, Cape Town 7700, South Africa
chttad005@myuct.ac.za, gt.mwalemba@uct.ac.za

Abstract. Technology is rapidly changing the landscape of the workforce and society. Business owners and other employers are actively seeking a labour force with the required digital skills and qualifications. Recognition of prior learning (RPL) has been identified as one of the means of upskilling the workforce and expanding access to formal qualifications for working adults. This research explores the motivations and experiences of professionals in technology-related fields seeking access to tertiary-level qualifications through the RPL route. Findings point to RPL fulfilling its intended role of being a necessity and, in some cases, the only means for experienced professionals without formal education to advance their skills and obtain formal qualifications necessary for their career advancement. There is also a need for higher learning institutions to extend their accessibility through similar programs and initiatives.

Keywords: Recognition of Prior Learning · ICT Industry · Skills Development

1 Introduction

South Africa's higher education sector is confronted with various challenges, some of which can be attributed to the past historically biased educational policies developed in the apartheid era. One of the key challenges is the unequal access to educational resources and learning opportunities. The National Qualification Framework (NQF) developed the recognition of prior learning (RPL) process in an effort to address the aforementioned inequities in higher education [11]. RPL provides an alternative means for those working in various sectors without formal qualifications to gain access to university studies based on their work experience. Implementing the RPL programme is part of the national commitment to redress inequities and support lifelong learning through creating more platforms and access to adult learners in South African academic institutions.

To date, many higher institutions within South Africa have implemented RPL as one of the official channels to gain access to tertiary qualifications. This study aims to explore the implementation of RPL within the information system (IS) field. Specifically, the

H. E. Van Rensburg et al. (Eds.): SACLA 2023, CCIS 1862, pp. 152–163, 2024.
https://doi.org/10.1007/978-3-031-48536-7_11

study seeks to answer the following research question: *What are the motivations and experiences of IS professionals pursuing tertiary-level qualifications through the RPL programme?* In asking this question, the study seeks to understand the perspectives and experiences of adult learners who have been admitted to pursue ICT-related postgraduate qualification via the RPL route. Although the study focuses on the experiences of individuals studying in an Information Systems department at a South African university, the findings of this study have the potential to contribute to ongoing discussions on subjects such as the role of adult education and life-long learning. Most importantly, the study can help to shed light on the role of RPL in addressing inequalities and expanding access to education, especially in the ICT field, which is acknowledged to have persistent skills shortages.

2 Literature Review

2.1 Background

South Africa is characterised by a history of exclusion of the majority from education, training, and employment opportunities. To redress some of these hindrances, RPL was instituted [14] as a policy to help reform education and training. The Department of Basic Education (DBE) asserts that South Africa needs equal access to lifelong learning, training, and educational opportunities to contribute towards improving its quality of life and building a peaceful, prosperous, and democratic South Africa [7].

2.2 Origins of RPL

The concept of RPL is an opportunity for South Africa because it can be viewed as a response to economic and social demands for change. RPL serves to provide an equitable appreciation of the knowledge, skills and abilities gained in the workplace [11]. This stems from the growing awareness of learning from experience in various contexts, such as formal and informal work and should be subsequently acknowledged and rewarded by educational institutions [4]. RPL was motivated by social and political responsibilities to increase the participation of Black South Africans excluded from many occupations and quality education during the apartheid era. Thus, the RPL initiative can be associated with individual and social justice [4]. RPL is often used as a stepping stone to higher access to education, lifelong learning, and gaining credits towards desired qualifications.

Within the ICT sector, RPL is also a means to contribute to closing the digital skills gap. RPL is a phenomenon containing a variety of practices, contexts, and conceptions [1]. Where individuals may lack the pre-requisite requirements, entry requirements, or credits, RPL may allow access. The individual will be granted advanced standing or exemption into the programme based on their prior qualifications. RPL understands that knowledge is attained in many different contexts. Therefore, it can be articulated and assessed against academic regulations to be formally recognised [16]. The primary and fundamental assumption of RPL is that learning and attaining knowledge is not reserved for formal educational institutions alone. According to RPL, knowledge can be achieved through formal, informal, and non-formal constructs [9]. This requires an academic standard to recognise the knowledge from a global perspective at tertiary institutions. The working definition of RPL is adopted here in the tertiary education context, where definitions may vary [1].

2.3 Overview of RPL

In South Africa, the NQF mandates the South African Qualifications Authority (SAQA) to develop a National Policy for the Recognition of Prior Learning. SAQA strongly advocates RPL for redressing and transforming through higher education. SAQA recognises that RPL can be an effective mechanism to award full or partial credits towards qualifications [13]. The policy was created to work "collectively towards demonstrably changing the lives of RPL candidates, including workers and learners of all ages, unemployed people, and other marginalised groups" [22]. Types of RPL may include:

RPL for Access. Recognising higher learning that provides alternative access to an academic program for individuals who do not fit the conventional entry criteria for admission. The RPL for access applies to prior learning and experience gained by an accredited institution or work-based training provider. SAQA [21] defines access as the chance to seek education and training, including applicable degrees, partial degrees, professional titles, job prospects, and career advancement. When it comes to RPL for access, it is believed that there are people with the acquired skills, common knowledge and work experience that lack relevant certification [23].

RPL for Credits. Acknowledging that a learner has sufficiently mastered specific course content through prior means and awarding credits for or towards a qualification. SAQA [21] describes credit as the amount of learning necessary to acquire a qualification, measured in notional study hours, in order to attain a qualification. This would imply that for every unit a person studies, there are a number of credit points earned towards the outcome [23]. In relation to RPL, an individual can claim and be evaluated for specific credits for their previous knowledge. These credits will then be transferred towards a learning programme. This principle usually applies to workers who are unemployed but have acquired the relevant skills. Credit transfer means the reallocation of credits towards a part qualification or qualification recognised by the NQF framework [21]. This is the RPL type that receives focus here.

2.4 RPL Process and Purpose

The RPL process varies between countries, where the core of RPL involves a few key activities. Typically, the RPL application process involves the candidate identifying the relevant institution and qualification, creating a portfolio of evidence of experience and submitting the portfolio together with a detailed CV to the institution for consideration. It's also possible that the RPL process can include the requirement that the candidate completes a set assessment. An appointed committee will deliberate on the per-performance of the candidate and whether they have met the requirements for admission whilst making a note to adhere to the appointed proportion of RPL students permitted. In this case, the South African Council on Higher Education (CHE) stipulates that the number of students admitted into any programme must not exceed 10% [6].

Consensus holds that RPL allows those who have acquired important information and skills acquired outside of traditional formal learning that can be acknowledged for credit toward a certificate, diploma or degree and used for gaining access to employment and other possibilities. RPL is, however, defined somewhat differently between nations. It can be a way of recognising knowledge and work done to acquire entry into higher education or to obtain qualification credits [10]. In the United States, it is more frequently utilised for credit towards a degree in the higher education sector than for access. Whilst in Australia, RPL, in contrast, focuses on the Vocational Education and Training sector, where it involves credit transfer and articulating agreements, rather than the higher education sector specifically [10].

As for South Africa, equity would be one of the primary reasons for RPL in New Zealand and Namibia [19, 25]. Although other uses for RPL are starting to emerge internationally, its original goals were to increase access to higher education and training and to recognise knowledge and abilities for academic credit. These goals demonstrate how RPL addresses social, political, and economic imperatives, just like in South Africa, and are not confined to a narrow academic focus [10, 12].

2.5 RPL Implementation

Whilst South Africa has achieved political freedom, economic freedom remains largely out of reach. At the core of this challenge lies the critical underdevelopment of professional skill sets in disadvantaged communities, where formal education continues to define the perceptions, mindsets, and expectations of how success ought to be calibrated. To this day, only a minority of the population has the privilege of being able to complete formal studies [13]. In such a context, alternative educational pathways become of critical importance for excluded individuals, hence the need to develop and promote RPL as part of an inclusive educational system.

Implementing RPL is the responsibility of various structures, stakeholders, boards, institutions, and the candidate. Generally, these structures play different roles in advancing the implementation and effectiveness of a given RPL programme. In accordance with national RPL policies, quality assurance for RPL processes is required. RPL records must be submitted by providers to the National Learners' Records Database (NLRD) for study in strict confidence to identify trends. Factors that have been identified as being pertinent to the effective implementation of the RPL include funding, quality assurance, effective delivery/alignment of RPL and rights and responsibilities of all role-players implementing/experiencing RPL [13, 21].

2.6 Experiences from RPL Implementation in Various Disciplines

There have been many implementations of RPL (or a close version of it) across different countries and disciplines. Research points to successful RPL implementation in nursing not only in South Africa but also in other countries such as Australia, the United Kingdom, Netherlands [15, 17, 18, 24]. Similarly, RPL has been implemented in various other disciplines, such as education and/or teacher training programmes, hospitality, and even agriculture to name a few [2, 20]. Studies report mixed success of RPL programmes, with some referencing very positive outcomes of the programme Donoghue et al. [8] and others pointing to RPL implementations riddled with challenged and negative student experiences [24]. Cooper and Harris [5] suggested that it cannot be taken for granted that students may easily convert their experiential knowledge to academic knowledge by simple reflection. Prior knowledge, which frequently takes the shape of established practices/habits, has a significant impact but can also act as a barrier to learning and transformation [26]. However, for the most part, there is an acknowledgement of the important role and contribution offered by the RPL programme when implemented effectively. There is also an emphasis on the need to understand the RPL candidate's profiles and experiences so that proper interventions can be put in place to manage their transition into the academic environment. That said, although South African institutions are increasingly offering IS and technology-related programs at various levels, there's still limited research that explores the incorporation of RPL into these programmes. This study represents such effort by specifically focusing on exploring the experiences of RPL candidates in IS postgraduate programmes.

3   Research Method

The research was conducted as a case study focusing on the Department of Information Systems at a South African university. The Information Systems Department offered part-time courses which allowed students entry through the RPL. The rotation included Information Systems Management, Business Analysis and Systems Analysis, as well as Cybersecurity. After successfully completing these courses, candidates proceed to the second year, where they can qualify with an honours degree. A table with a brief description of the participants who were identified using purposive sampling can be found in Tables 1 and 2. Data was collected through semi-structured interviews making use of mostly open-ended questions. Once data was collected, the transcripts were added to NVivo for data analysis and initial themes were derived inductively through thematic analysis [3].

Table 1. RPL Staff and Alumni Participant

	Title	Role
1	Staff01	Course Convenor, Admissions Committee
2	Staff02	Course Convenor, Admissions Committee
3	Staff03	Head of Department
4	Staff04	Administrator
5	Staff05	Course Convenor, Admissions Committee
6	Alumni01	IT Support Technician
7	Alumni02	Systems Development Manager
8	Alumni03	Software Development Manager

Table 2. RPL Student-Staff and Student Participants

	Title	Course	Role
9	Student-Staff01	Information Systems Management	Help Desk Technician
10	Student01	Business Analysis & Systems Analysis	Business Analyst
11	Student02	Information Systems Management	Unemployed (Past: Hotel Operations Intern)
12	Student03	Information Systems Management	Data Engineer
13	Student04	Cybersecurity	IT Support Specialist, Project Manager
14	Student05	Cybersecurity	SecOps Engineer

4   Research Findings

4.1   Overview of Findings

Figure 1 displays visual representations of key findings and major themes derived from thematic analysis. The findings presented as part of this study are grouped into two key aggregate dimensions/themes namely 'motivations' and 'experiences'.

4.2   Motivations

This study's 'motivations' theme covers students' reasons and intentions when joining IS postgraduate programmes at university through the RPL route. As these are mature students with various backgrounds, often there is a purpose when coming to university.

Career Advancement/Promotion and Industry Requirements
One of the emerging themes derived from candidates pursuing RPL revolves around the desire to enhance their careers. Respondents attributed their drive to pursue postgraduate qualification through RPL as a means to attain promotions or change careers to those they perceived as carrying more potential.

"Myself and one other chap that was on the programme with me. We got promotions shortly after getting our degrees. So that was something I don't believe either of us would have gotten without a degree." (Alumni02).

This was also emphasised by a staff member who has admitted and taught several RPL students over the years.

"many of them have hit a ceiling cause they haven't got a qualification, so the minute they get that qualification or just the fact that there's the threat of the qualification, they move up or they move jobs." (Staff02).

Difficulties in attaining a promotion have been seen when students do not have the formal qualifications deemed necessary for certain roles. As expressed by a staff member,

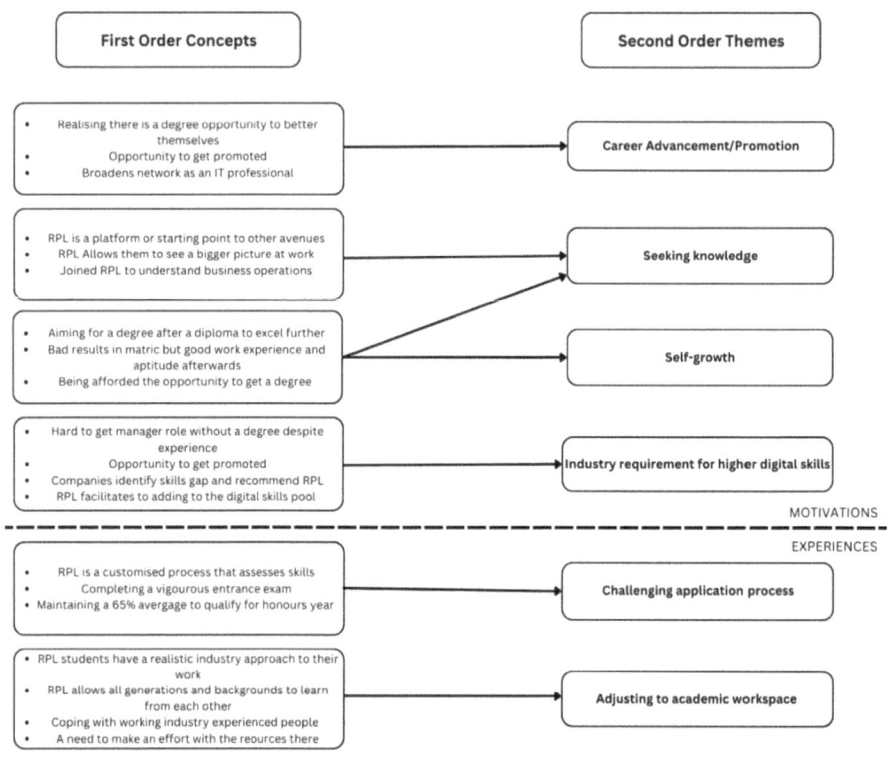

Fig. 1. RPL first-order concepts and themes emerged

"They've got to the top of their field. They offered in fairly senior positions, but they can't get any higher because they haven't got a degree and so they are incredibly motivated and very experienced and mature."

This is also echoed by one of the students.

"Like, it's hard to get a job as a manager because you don't have your degree, but you have 20 years experience, so they want you. But once you have a degree with the experience, they will immediately take you over somebody that has a degree without any experience." *(Student04).*

Seeking Knowledge

For some, the pursuit of additional skills or knowledge was their main motivation to study via the RPL route. They believe studying at a higher learning institution, especially those well recognised and rated globally, will enable them to either attain new skills or enhance their existing ones.

"I think it was in line with my career goals and where I wanted to see myself on upskilling, especially adding on to like my technical side of skills and yeah, I think I very much aligned with the programme, and it helped me prepare for my career and I thought it was worth it." *(Student01).*

This shows that within the concept of upskilling, there is an opportunity to think differently in the industry after having attended the academic institution.

"...It will help me in the future. I think it has definitely taught me to be a bit more. Yeah, maybe a little bit more strategic, thinking about how things are done, I feel more confident in what I do after having done the degree." (Alumni03).

Self-Growth

Some respondents were driven by self-growth. We separated seeking knowledge and self-growth sub-themes as follows: Seeking knowledge is about the pursuit of specific job-related skills, whereas self-growth is about developing oneself as a whole person independently of specific job-related needs. Below are a few examples of self-growth as motivation.

"It really helped me with my critical skills. It really played a part in my I'd say self-reflection." (Student01).

"It's one of our open requirements in my workplace. I understand it better now. I see the full picture now and it makes me question how many people in my organisation are amongst my peers. I understand that big picture." (Student04).

4.3   Experiences

The aggregate dimension of experiences refers to the series of events that occurred during the students' time in the institution. This includes entering the university hemisphere and their experience during their time.

Challenging Application Process

As the RPL programme provides university entry without undergraduate qualifications, a means of skills assessment is necessary. Applicants must submit a curriculum vitae (CV), a portfolio, motivational letters, and any other evidence showcasing their skills and experiences as part of the application process. They were also expected to write and pass an entrance exam. Responses show that some found the application process, especially the assessment, challenging.

"I don't think any of us said anything for 15 min at first and then one of us popped up and said I failed and then the other two, so we definitely failed. So we were adamant, actually, all three of us that went down together were adamant that we had failed that test." (Alumni02).

"I know that we had to do an assessment in order to qualify for the programme further. So I think that assessment really tested me. Integrating your practical as well as theoretical knowledge and application-based questions especially and it wasn't only focused on the area that I was going to pursue the course in, it was an all-rounded assessment." (Student01).

These students' experiences were also echoed by staff members involved in the application process. The staff member also adds insights on how the challenging application process shapes the mindset of students who have been through the RPL process compared to those accepted via traditional entry requirements.

"...so whereas the other students just come along nonchalantly, the RPLs have been through this traumatic entrance exam. So they feel, they are aware that they're different.

They're aware that everyone else has got a degree and they haven't and they've come through this test for an exam and they therefore think they have to work a bit harder and then they tend to, and then they do better." (Staff02).

Adjusting to the Academic Environment

This theme speaks to student experiences of the RPL programme. As staff are also involved in the process, they weigh in on assessing how RPL students cope in the academic environment. RPL students are seen as people with a "realistic" approach, as the majority of them come from industry.

"Often they've got a realistic idea what's at the industry. They can do presentations very well. Again, their writing is often very marketing oriented rather than factual oriented and the same for the type of information sources that they use in especially people" (Staff01).

From the student's perspective, the programme presented a challenge in terms of time management.

"Cons to it I'd say the only thing that was kind of something that I wasn't really prepared for was how much time it required in terms of it was time-consuming throughout it. So, uh, scheduling was an important part for me throughout the whole programme." (Student01).

There were also reflections on the need for cultivating a different academic mindset (different from the workplace mindset) that's seen as necessary in order to be able to write and communicate in an academic style.

"But there is a lot that you need to pick up along the way. That's going to make you better and help you to cope and things like that. I think things like the writing centre are a big deal because if you struggle. If you're not in that mindset and you can't figure out how to write an academic paper then you know that was the place to go but you needed to make the effort." (Alumni01).

5   Discussion

The findings point to an alignment between the intended goals of setting up RPL initiatives and the experiences of students who have been part of the programme. According to the study, the RPL option offers access to different programmes within the ICT field and, subsequently, an opportunity to obtain qualifications that can advance their career aspirations as well as self-growth. It's also a great opportunity to enhance their expertise by combining university knowledge with their industry experience. It is noted that prior knowledge can be a powerful influence but can also act as a barrier to learning and change; therefore, there is a need to strike a balance between the two [26].

Candidates pursuing the RPL come from various learning backgrounds. The findings highlight challenges in adapting to the academic environment specifically the academic mindset and writing style. Again, this is not unique to IT programmes as it is also widely reported in the literature. The common solution to this challenge, according to the literature, requires the provision of additional resources and customised support to RPL candidates [24]. However, this was not highlighted in this study. This could be attributed to the observation that the stated challenges didn't seem to have a significant negative impact on the overall successful outcome of the programme. There's also the acknowledgement that the number of RPL candidates is still capped at a lower percentage which could mean only very strong candidates get through [6].

6   Conclusion

South Africa is a country with rich institutional and experiential knowledge but palpable inequities of access. RPL has been introduced to the national educational system as part of a necessary redress. RPL proposes, in this context, a combination of experiential and academic learning, where people without a formalised qualification are recognised for the years of experience they have attained in the industry. An exploratory case study was conducted to understand the motivations and experiences of IS professionals who have taken part in the RPL process. Results obtained were grouped under two overarching themes of motivations and experiences.

The motivations of RPL students varied, but all came down to seizing the opportunity of access to additional knowledge and formal qualifications. Both the additional knowledge as well as formalised qualification were identified as instrumental for career advancement at the workplace. This meant RPL was able to give candidates career opportunities that they would have otherwise not been able to access. This, for the most part, aligns with the intended objective of the RPL initiative. However, students also highlighted the challenges that come with the application process and adjusting to the academic environment.

The findings of this study point to the necessity for universities to continue expanding access through similar initiatives. Currently, the government has restricted the percentage of students in a class that can be admitted via the RPL process. The many positive experiences of the initiative highlighted in this study hint at the need to reconsider such limitations in order to make the RPL initiative more accessible. This is especially important in the ICT field, which is well known for its persistent skills shortage and lack of diversity, especially in senior positions.

One of the key limitations of this study is its sample size. Success at one institution isn't necessarily an indication of the overall success of the initiative. There's a need for further studies that will look at RPL implementation in other institutions. There's also a need for other aspects of the programme, such as RPL students' performance relative to those admitted through traditional entry requirements. Lastly, not all students who apply to join RPL are successful. Further studies can explore the reasons, experiences and perceptions of those whose applications are denied.

Acknowledgements. All the students, staff and alumni who contributed to make this study possible.

References

1. Andersson, P., et al.: Introducing research on recognition of prior learning. Int. J. Lifelong Educ. **32**(4), 405–411 (2013). https://doi.org/10.1080/02601370.2013.778069

2. Baumeler, C., et al.: Recognition of prior learning in professional education from an organisational perspective. Int. J. Lifelong Educ. **42**(2), 208–221 (2023). https://doi.org/10.1080/02601370.2023.2177759

3. Braun, V., Clarke, V.: Using thematic analysis in psychology. Qual. Res. Psychol. **3**(2), 77–101 (2006). https://doi.org/10.1191/1478088706QP063OA

4. Castle, J., Attwood, G.: Recognition of Prior Learning (RPL) for access or credit? Problematic issues in a university adult education department in South Africa. Stud. Educ. Adults **33**(1), 60–72 (2001). https://doi.org/10.1080/02660830.2001.11661441

5. Cooper, L., Harris, J.: Recognition of prior learning: exploring the 'knowledge question.' Int. J. Lifelong Educ. **32**(4), 447–463 (2013). https://doi.org/10.1080/02601370.2013.778072

6. Council on Higher Education (CHE): Recognition of Prior Learning, Credit Accumulation and Transfer, and Assessment (2016)

7. Department of Basic Education (DBE): Department of Basic Education (DBE) – Overview. https://nationalgovernment.co.za/units/view/7/department-of-basic-education-dbe#:%7E:text=The%20vision%20of%20the%20Department,prosperous%20and%20democratic%20South%20Africa. Last accessed 7 Apr 2023

8. Donoghue, J., et al.: Recognition of prior learning as university entry criteria is successful in postgraduate nursing students. Innov. Educ. Teach. Int. **39**(1), 54–62 (2010). https://doi.org/10.1080/13558000110102896

9. Hendricks, M.N.: The recognition of prior learning in higher education: the case of the University of the Western Cape (2001)

10. Hlongwane: Recognition of Prior Learning (RPL) implementation in library and information science (LIS) schools in South Africa (2014)

11. Hlongwane, I.: Recognition of prior learning implementation in library and information science schools in South Africa: a literature review. Africa Educ. Rev. **15**(3), 113–129 (2018). https://doi.org/10.1080/18146627.2017.1353396

12. International Labour Organization (ILO): International Labour Organization Recognition of Prior Learning (RPL) Learning Package, Geneva (2018)

13. Luckan, Y.: Introduction. In: Luckan, Y. (ed.) The Recognition of Prior Learning in Post-Apartheid South Africa: An Alternative Pedagogy for Transformation of the Built Environment Professions, pp. 1–130. Routledge (2021). https://doi.org/10.4324/9781003121428-1

14. McIntyre, G.: How Rpl Empowers South African Businesses. http://www.hrpulse.co.za/editors-pick/234025-how-rpl-empowers-south-african-businesses

15. Muller, J., et al.: Recognising the skills and competencies of non-EU foreign nationals: a case study of the healthcare sector in the Netherlands. Soc. Policy Soc. **16**(4), 681–691 (2017). https://doi.org/10.1017/S1474746417000264

16. Nel, H.: A Holistic Approach to the Recognition of Prior Learning. Presented at the April (2018)

17. Pryor, C.S.: GCert Hlth(Clinical Forensic Nursing) MRCNA Central Northern Adelaide Health Service. Australian J. Adv. Nursing **30**(2), 40–47

18. Scott, I.: Accreditation of prior learning in pre-registration nursing programmes 2: the influence of prior qualifications on perceived learning during the foundation year. Nurse Educ. Today **30**(5), 438–442 (2010). https://doi.org/10.1016/J.NEDT.2009.10.002

19. Shaketange, L.L.: Challenges and opportunities for implementing recognition of prior learning at the University of Namibia. Creat. Educ. **9**(13), 2070–2087 (2018). https://doi.org/10.4236/CE.2018.913150

20. Snyman, M., van den Berg, G.: Experiences of nontraditional students and academics of the recognition of prior learning process for admission to graduate studies: a South African case study in open distance learning. J. Continuing High. Educ. **70**(2), 71–87 (2022). https://doi.org/10.1080/07377363.2020.1861577

21. South African Qualifications Authority (SAQA): SAQA Bulletin. 19, 1, (2020)
22. South African Qualifications Authority (SAQA): The South African Qualifications Authority National Policy for the Implementation of the Recognition of Prior Learning. https://www.saqa.org.za/wp-content/uploads/2023/02/Updated-RPL-Policy-website-version-Resized.pdf. Last accessed 15 June 2023
23. Thobejane, D.V.: Investigation into the challenges for an implementation of recognition of prior learning in further education and training, in Limpopo Province. University Of Limpopo (2016)
24. Udeagha, G.M., et al.: Experiences of students who gained entry to a nursing college through recognition of prior learning: A phenomenological study. Nurse Educ. Today **117**, 105474 (2022). https://doi.org/10.1016/j.nedt.2022.105474
25. UNESCO: New Zealand RVA country profile in education and training. https://uil.unesco.org/document/new-zealand-rva-country-profile-education-and-training (2014)
26. Valentine, B., et al.: A developmental approach to recognition of prior learning in social work field education. Aust. Soc. Work. **69**(4), 495–502 (2016). https://doi.org/10.1080/0312407X.2016.1168462

First Year Computing Students' Career Choice Influencers

Margaret Cullen[1] , Andre P. Calitz[2](✉) , Malibongwe Twani[2] ,
and Jean Greyling[2]

[1] Nelson Mandela University Business School, Port Elizabeth, South Africa
`Margaret.Cullen@Mandela.ac.za`
[2] Department of Computing Sciences, Nelson Mandela University, Port Elizabeth, South Africa
`{Andre.Calitz,Malibongwe.Twani,Jean.Greyling}@Mandela.ac.za`

Abstract. Research indicates that first year students who have chosen a career in Computer Science (CS), Information Systems (IS), Information Technology (IT) and other related computing subjects were generally influenced by parents, teachers, and role models. Current research indicates that exposure to new technologies, gaming and programming Apps at school level can influence a scholar's IT career choice. Theories relating to career choice have focused on the characteristics of individuals and their environment. Presently, CS/IS/IT departments do not know who or what influenced first year students to pursue a career in IT. Understanding first year students' academic IT career choices influencers would assist academic departments to improve methods and strategies to recruit first year CS, IS and IT students.

The purpose of this study was to determine who or what influenced current registered first year students to choose an IT career. A questionnaire was completed by first year CS, IS and IT students at a Comprehensive University. The findings contradict the literature, which states that parents and teachers, influence a first-year student's IT career choice. Thirty-five percent of the respondents did not know their farther, 19% did not know their mother and 38% did not know their father's occupation. Social media and IT role models were important influencers for first year CS/IS/IT students speaking an African language at home. This study introduces an innovative gaming project introduced at schools in order to introduce school children, parents and teachers to coding.

Keywords: IT career choice · influencers · career advice

1 Introduction

Influence is defined by the Merriam Webster Dictionary as "the power or capacity of causing an effect in indirect or intangible ways", whereas influencer is defined as "one who exerts influence, who inspires or guides the actions of others". A person's career choice is influenced by several factors, which results in a process of defining what they want to do and exploring the available options [1]. Career choice is not determined by one single factor, rather it is determined by the interplay of several social, cultural

H. E. Van Rensburg et al. (Eds.): SACLA 2023, CCIS 1862, pp. 164–179, 2024.
https://doi.org/10.1007/978-3-031-48536-7_12

and economic factors [2]. Research has been conducted globally on the factors that influence career choice with the following factors generally being identified: family influence, specifically parents, teachers, passion, values, a sense of belonging and self-efficacy [3]. The career choice theories assume that the individual has knowledge of the career options available.

Siddiky and Akter's [2] study in Bangladesh, a developing country, indicated that career development training plays an important role in developing the competencies of the students for jobs, however most of the students do not have access to such training. In addition, they report that universities in Bangladesh do not have career guidance or counselling programmes. The process of making a career choice is complex and unique for everyone depending on cognitive factors and social structures of the individual's environment [4, 5]. Parental education and careers, parental encouragement and advice are critical in students' choices of careers [5].

First year students who have chosen a career in Computer Science (CS), Information Systems (IS), Information Technology (IT) or other computing related fields of study were generally influenced by people and events they had contact with during their daily lives [6]. Mtemeri's [5] study proposed a career guidance model with the following six components: training career guidance teachers, planning career guidance activities, availing adequate resources, training parents, peer education on career guidance and linking students with industry.

The following academic theories relating to career choice influencers were reviewed. Two academic theories on influence were consulted. Kelman's Social Influence Theory [7] proposes that an individual's attitudes and beliefs and resulting behaviours and actions are influenced by others through compliance, identification and internalisation. Latane's [8] Social Impact Theory suggests that the greatest influence is the action of others, in that others can persuade, inhibit, threaten or support. These two theories are relevant to the decisions students make when deciding on a career path. Influencers are important to link students to careers [2].

Krumboltz Social Learning Theory of Career Decision Making (SLTCDM)[9], proposes that people make career choices through a series of planned and unplanned learning opportunities arising from an individual's social environment. The theory emphasises four beliefs:

- Beliefs about our own abilities, which influence our future planning are known as genetic endowment. These include gender, race, intelligence and special abilities, which are inherited and built in [2];
- Beliefs about our environment or context, upon which assumptions about the future are made. These include the effects of family, teachers and resources, technology, training opportunities, occupational factors, labour market and are conditions usually beyond the individual's control affecting a person's career decision-making [2];
- Beliefs around the skills the student thinks they have, whether innate or learned. Learning experiences involve instrumental learning and associative learning that have effects on career choice [2]; and
- Actions taken based on the above, because of what was learnt. This includes task approach skills like learning skills, goal setting and obtaining career information [2, 10].

Literature indicated that students' career choice is generally influenced by parents, teachers, career counsellors and role models [2, 6]. The occupation, education and advice of a child's parents are influencers in students' choices of careers [5]. Social media also has an influence on a student's career choice [11]. Other interventions, such as computer clubs, gaming and mobile Apps, workshops and IT camps have created IT career awareness amongst children [12, 13]. In South Africa, the TANKS coding app, which includes coding kits, lesson plans and teacher training material has reached over 100 000 scholars and Tangible Africa has trained over 20 000 teachers [13]. Limited research has been conducted on CS/IS/IT first year students' career choice influencers in South Africa.

The layout of the paper is as follows. In Sect. 2, the research problem and research questions being investigated are introduced. Section 3 provides a literature review, which focuses on a systematic literature review on the factors that affect career choice decisions. The results are discussed in Sect. 4. An innovative case study is introduced in Sect. 5. Section 6 presents the conclusions where limitations and future research are also presented.

2   The Research Problem

First year students who have made a CS/IS/IT career choice were influenced by people, events or formal and social interactions. The problem investigated in this study is that CS/IS/IT and other computing departments do not know who or what influenced their first-year students to choose an IT career. The research questions addressed in this study were:

1. *What theories apply to career choice influencers?*
2. *Who influenced the first year CS/IS/IT students to pursue a career in IT?*
3. *What other career choice influencers did first year students receive?*
4. *How did IT role models influence first year students' IT career choice?*
5. *How can innovative approaches, which influence IT as a career choice, assist scholars to make career decisions?*

3   The Research Design

A first-year questionnaire was compiled based on a similar questionnaire used in previous studies [14]. In order to determine personal perceptions and honest information, it was decided to keep the survey anonymous. The first-year questionnaire included the following sections:

- Biographical details;
- Sources of influence; and
- IT role models.

The questionnaire was captured using the on-line survey tool, QuestionPro. The data were collected from first-year students enrolled for CS, IS, IT degree and diploma programmes at a comprehensive university in South Africa. The data were statistically analysed using Statistica, with the assistance of the university statistical consultant. Ethics approval was obtained from the University Ethics Committee.

4   Literature Review

A systematic literature review was conducted for this study. A Systematic Literature Review (SLR) is defined as rigorous, methodical and transparent in the methods used to identify the research included in a study [15]. Kitchenham et al. [16] further note the advantages of an SLR are: 1) it summarises existing evidence; 2) identifies limitations and benefits in the current literature; 3) identifies gaps and areas of further investigation, and 4) provides a framework to position new research activities. The SLR covered literature on the factors that influence career choice. Inclusion criteria ensured that articles that 1) investigated factors that influence decision making at university and college, and 2) studies discussing theories on decision making. The inclusion and exclusion criteria resulted in 28 articles, which the researchers felt was too low. Kitchenham et al. [16] recommend a manual search of the literature checking references from relevant articles, which is called a backward search. Therefore, a backward process was undertaken that increased the final list to 39 articles.

4.1   Background Information

The SLR results indicate that background information included gender and race as important factors. Race and gender were included in most of the SLR studies, with a combined count of 31 out of 39 articles. The factors age, disabilities and years of study were included, as they influence students' career choice. Therefore, twenty studies use gender as a factor. Matthew et al. [17] state that gender can influence the type of careers students pursue. In support, Bock et al. [18] included gender as one of the factors that affect a person's decision to choose a career.

Matthew et al. [17] in their study examining students' interest in accounting courses, identified that the age of students can determine the choice of career. However, they do admit that it could be that the age range was not large. On the contrary, Aivaloglou and Hermans [19] in their study, that examined the Hispanic students' interest in STEM fields, found that age did not affect students' career orientation.

Elias and Brian [20] state that a few studies examine racial and ethnic groups career decision influencers. The SLR results support Mein et al. [21] in a study that examined Hispanic undergraduates and their choice to study engineering. Mein et al.'s [21] results indicate that ethnicity plays a role in career choice, as it highlighted different challenges amongst races.

4.2   Parents, Family and Friends

The SLR results indicated 26 articles out of 39 articles in total applied family, friends, teachers and friends as influencers of career choices. The influences were either positive or negative. The prominent influencers were family, which includes parents and teachers and career guides with a combined occurrence of 28. Parents and families were very influential in children's' career decisions [3, 6]. Friends, mentors and role models were the second-largest occurring theme [3, 22, 23].

4.3 Sources of Information

Downes and Looker [24] discovered that parental education was a key influencer of career choices. Their study examined factors that contribute to low participation in computing and information technology courses at secondary schools. In support, Mein et al. [21] noted that students whose parents were engineers, were involved in engineering talk and activities from a very young age. Therefore, indicating the influence that parents have on their children. Mein et al. [21] further noted positive school-based support from teachers and mentors also plays a positive role in guiding students to STEM fields.

Govender and Khumalo [25] indicated that a lack of knowledge by influencers, such as family and friends led to students not choosing to pursue IS studies. Additionally, Seymour and Serumola [26] found that students' decisions to enrol or not to enrol for IS-related courses were influenced by the lack of information from their teachers. Lee et al. [11] note that social media has changed the way people and organisations share information. Additionally, input from social media was considered important in influencing a student's career choice.

4.4 Learning Experiences

Out of the 39 articles of the SLR, only 18 articles examined learning experiences. Interest in school subjects (Maths and Science) is important for students' career choice. Early exposure to computer clubs and programming is important as six studies have shown that interest at school is a key factor for students' career decisions. Lastly, computer education, coding and problem-solving were key as a learning experience that has influences on students' career decisions [13]. Mein et al. [21] state that interest is important for involvement in computer projects and programming from a young age and studying experiences from school and teachers are helpful.

Limited awareness of computer careers does impact final career decisions. However, academic exposure to computer careers is important to generate interest and curiosity in the field and provide a broader picture. Interventions, such as computer clubs, gaming and mobile Apps, IT camps and hackathons, which could provide students with opportunities of learning different computing disciplines and understand various computing careers to increase awareness and interest in the computer field [12, 13].

5 Results

The qualitative and quantitative data analyses were conducted and included Exploratory Factor Analysis (EFA) and ANOVA statistics. EFA was used as it reduces data to a smaller set of factors and identifies the structure of the relationship between the factors. The research findings are discussed in the following sub-sections, namely:

- Biographical details;
- Sources of influence; and
- IT role models.

5.1 Biographical Details

The first-year survey was completed by 205 students who were registered for the BSc CS, BCom CS&IS, BIT (n = 92, 43%) and IT Diploma programmes (n = 113, 55%) in the Department of Computing Sciences and the School of IT at the XX University. Table 1 shows that the sample consisted of 154 males and 51 females. The sample included 43% Black, 48% White, 7%, Coloured and 2% Asian students. The citizenship of the total group was mainly South African (86%). A small number of students (14%) were from Botswana, Malawi, Namibia, Zambia and Zimbabwe. The home language spoken was Afrikaans/English (24%), Xhosa (43%), Zulu, including Sesotho, Tswana, etc. (18%) and other languages, such as Sepedi and Xitsonga (14%). The Home languages were finally categorised into two groups, namely Afrikaans/English and African.

Table 1. Demographics

Home language	Afrikaans/English	Xhosa	Zulu etc.	Other
	50 (24%)	88 (43%)	38 (18%)	29 (14%)
Age	**18 years**	**19 years**	**20-21 years**	**22+ years**
	53 (26%)	55 (27%)	55 (27%)	42 (20%)
Race	**Asian**	**Black**	**Coloured**	**White**
	5 (2%)	89 (43%)	19 (7%)	92 (48%)
Programme	**CS/IS/IT degree**	**Diploma IT**		
	92 (45%)	113 (55%)		
Nationality	**Foreigners**	**South African**		
	20 (14%)	185 (86%)		
Gender	**Male**	**Female**		
	154 (82%)	51 (18%)		

The career choice question was an open-ended question in which respondents had to indicate the career they want to pursue, after completing their studies. The responses were coded as indicated in Table 2. Into two IT job title categories, CS related and IS related.

Table 2. Job titles specified

Job title	n (%)
Software Developer, Programmer, Game Designer, Graphic Designer, Machine Learning and AI	110 (54%)
Business Intelligence, Business Analyst, Software Engineer, IT Manager, IT Specialist, System Analyst, UX/UI Designer	84 (41%)
Not specified	11 (5%)
Total	205 (100%)

5.2 Parents as Career Influencers

Children of parents who discussed university more frequently were more likely to attend university after completing their high school education [6]. Vernon and Drane [6] found that students from low socio-economic status backgrounds reported low expectations of attending university. They further found that students discussed career and academic pathways more with parents than with teachers and career counsellors.

The analysis of the responses first-year students provided regarding the occupation and qualifications of their fathers, were depressing (Table 3). Thirty-five percent of the respondents indicated that their father was unknown and 19% their mother. The analysis of the responses included responses, such as 'I don't know my father', 'I don't have a father', 'I don't know my dad!', 'null', 'dead' and 'n/a'. Seventeen students have lost one or both parents. Thirty percent of the respondents' mothers were unemployed and 27 students indicated that both parents were unemployed.

Table 3. Parents occupation and qualifications.

Occupation	Father	Mother	Qualification	Father	Mother
Professional	6 (3%)	4 (2%)	Less than Matric	45 (22%)	62 (30%)
Entrepreneur	14 (7%)	10 (5%)	Matric	33 (16%)	47 (23%)
White Collar	33 (16%)	39 (19%)	Higher Certificate	2 (1%)	4 (2%)
Pink Collar	2 (1%)	14 (7%)	Diploma/Advanced Certificate	18 (9%)	31 (15%)
Blue Collar	33 (16%)	33 (16%)	Bachelor's Degree/Advance Diploma	14 (7%)	23 (11%)
Unemployed	43 (21%)	62 (30%)	Postgraduate Degree	14 (7%)	16 (8%)
Pensioner	2 (1%)	4 (2%)			
Unknown	72 (35%)	39 (19%)	Unknown	78 (38%)	27 (13%)
Total	205 (100%)	205 (100%)	Total	205 (100%)	205 (100%)

Thirty-eight percent of the first-year students did not know their father's occupation. Fifty-three percent of the first-year student's mothers had a matric or less than matric qualification and thirty-six percent of their mothers had a post-matric qualification. Only 24% of the fathers had a post-matric qualification.

5.3 Sources of Career Influence

The career influence factor used a 3-point Likert scale and investigated the career influencers and to what extent they influenced the respondent's chosen career. Table 4 indicates that the Internet helped 60% of the respondents and role models were evenly divided between having an influence 40% and 38% not influenced at all. The source of career influence for first year students shows that family (31%), friends (18%), teachers (26%) and Career counsellors (29%) helped 'a lot' with career decisions. The Internet (60%) was the most useful source of information.

Table 4. Career advice influencers

Career advice	Not at all	A little	A lot
Family	68 (33%)	74 (36%)	64 (31%)
Friends	96 (47%)	72 (35%)	37 (18%)
Teachers	82 (40%)	70 (34%)	53 (26%)
Career guidance advisors	78 (38%)	68 (34%)	59 (29%)
Role models	78 (38%)	45 (22%)	82 (40%)
My religious circle	146 (71%)	43 (21%)	16 (9%)
The Internet	29 (14%)	51 (25%)	125 (60%)
Social media	68 (34%)	62 (30%)	76 (37%)

5.4 Exploratory Factor Analysis (EFA)

The EFA evaluation for the factor *Career Choice Influencers* resulted in a two-factor model (Table 5), with a total of 57.2% total variance explained. Factor one was renamed *Career Choice Influencers – Personal* with 6 items and factor two named *Career Choice Influencers – Media* with 2 items. The item: Role models was retained in the factor Career Choice Influencers – Personal, as it had a higher factor loading of 0.488 in the factor.

Table 5. EFA Loadings (2 Factor Model) – Career Choice Influencers (n = 205; Minimum significant loading = .300)

Items	Factor 1	Factor 2
Teachers	**.777**	.098
Career guidance advisors/teachers	**.738**	.150
Family	**.708**	−.068
My religious circle	**.686**	.120
Friends	**.661**	.297
Role models	**.488**	**.312**
The Internet	.056	**.884**
Social media	.116	**.863**
Explained variance	2.81	1.76
% of Total variance	35.1%	22.0%
Total % of Variance Explained = 57.2%		

The EFA for the factor *IT Role Models* retained three items, explaining 48% of the variance (Table 6). The item: *I have IT role models* did not meeting the required minimum significant loading of .300.

Table 6. EFA Loadings (2 Factor Model) – IT Role Models. (n = 205; Minimum significant loading = .300)

Item	Factor 1
There are IT professionals in my family	**.899**
I have family working in the IT industry	**.896**
I have friends working in the IT industry	**.480**
~~I have IT role models~~	~~.260~~
Total % of Variance Explained = 47.7%	

The Cronbach's alpha coefficients for the three factors are presented in Table 7. The career choice influencers showed 'Good' reliability, however the factor IT Role Models rated poorly.

Table 7. Cronbach's alpha coefficients

Factor	Coefficient	Reliability
Career Choice Influencers – Personal	0.78	Good
Career Choice Influencers – Media	0.74	Good
IT Role Models	0.55	Poor

5.5 Career Choice Influencers (ANOVA Results)

The results in Table 8 show that *Home language* is statistically significant (p = 0,005) and practically significant (Cohen's d = 0.33). Therefore, this depicts a difference in mean values for the *Home Language* of first year students speaking Afrikaans and English and first-year students speaking African languages at home. The post-hoc results in Table 9 affirm the differences between respondents' home language. First-year students speaking Afrikaans and English at home were less influenced ($\mu_1 = 1.67$) by media compared to female respondents ($\mu_2 = 1.85$). The results indicate that first-year students who speak an African language at home, choice of career is influenced more by personal factors.

The results in Table 10 show that *Gender* is statistically significant (p < 0005) and practically significant (Cohen's d = 0.59). *Home language* is also statistically significant (p < 0005) and practically significant (Cohen's d = 0.56). Therefore, this depicts a difference in mean values for the *Gender* and *Home language* for first-year students. The post-hoc results in Table 11 affirm the differences between the respondent's gender and home language. Male respondents were less influenced ($\mu_1 = 2.00$) by media compared to female respondents ($\mu_2 = 2.40$). The results indicate that media has a greater influence on females' career choice. First year students speaking Afrikaans and English at home were less influenced ($\mu_1 = 1.96$) by media compared to first year students speaking a Black language at home ($\mu_2 = 2.34$).

Table 8. Univariate ANOVA Results – Career Choice Influencers – Personal

Effect	F-value	D.F	p-value	Cohen's d	Prac Sig
Gender	3,39	1; 227	,067	n/a	n/a
Age Category	0,36	1; 227	,550	n/a	n/a
Race	0,98	2; 227	,378	n/a	n/a
Home Language	*8,11*	*1; 401*	*,005*	*0,33*	*Small*
Father Highest Qualification	0,51	4; 227	,729	n/a	n/a
Mother Highest Qualification	1,28	4; 227	,278	n/a	n/a

Table 9. Post-hoc Results – Career Choice Influencers – Personal

Effect	Level 1	Level 2	μ_1	μ_2	p-value	Cohen's d	Prac Sig
Home Language	Afrikaans /English	African	1,67	1,85	,005	0,33	Small

Table 10. Univariate ANOVA Results – Career Choice Influencers – Media

Effect	F-value	D.F	p-value	Cohen's d	Prac Sig
Gender	*18,05*	*1; 227*	*< .0005*	*0,59*	*Medium*
Age Category	0,91	1; 227	,341	n/a	n/a
Race	1,95	2; 227	,145	n/a	n/a
Home Language	21,72	1; 401	< .0005	0,56	*Medium*
Father Highest Qualification	1,69	4; 227	,153	n/a	n/a
Mother Highest Qualification	0,25	4; 227	,908	n/a	n/a

Table 11. Post-hoc Results – Career Choice Influencers – Media

Effect	Level 1	Level 2	μ_1	μ_2	p-value	Cohen's d	Prac Sig
Gender	Male	Female	2,00	2,40	,000	0,59	Medium
Home Language	Afrikaans /English	African	1,96	2,34	,000	0,56	Medium

The results in Table 12 show that *Age Category* is statistically significant (p = 0,008) but not practical significant (Cohen's d = 0.14). *Home language* is also statistically significant (p < 0005) and practically significant (Cohen's d = 0.70). The post-hoc results (Table 13) confirm the differences between respondents whose Age Category is less than 21 years (μ_1 = 1.25) compared to those aged 21 plus years (μ_2 = 1.29). This indicates a marginal difference in the way *IT Role Models* were perceived by respondents younger than 21 years and those older. The post-hoc results (Table 13) further confirm the differences between respondents whose *Home language* is one of the African languages ($\mu_2 = 1.22$) compared to those with a home language Afrikaans and English ($\mu_1 = 1.41$). This indicates a difference in the way *IT Role Models* were perceived by respondents, with respondents having an African home language possibly being more influenced by IT role models.

Table 12. Univariate ANOVA Results – IT Role Models

Effect	F-value	D.F	p	Cohen's d	Prac Sig
Gender	0.02	1; 227	.903	n/a	n/a
Age Category	*7.05*	*1; 227*	*.008*	*0.14*	*n/a*
Home Language	*42.75*	*1; 227*	*<.0005*	*0.70*	*Medium*

Table 13. Post-hoc Results – IT Role Models

Effect	Level 1	Level 2	μ_1	μ_2	t-test p	Cohen's d
Age Category	<21 years	21+ years	1.25	1.29	.008	0.14
Home Language	Afrikaans/ English	African	1.41	1.22	.000	0.70

5.6 Tangible Africa Coding Project – A Possible Solution

The TANKS coding app was developed as an honors project at the Department of Computing Sciences at Nelson Mandela University [27]. As shown in Fig. 1, it makes use of "tokens" that learners' piece together to build code without a computer.

Fig. 1. TANKS app

TANKS has been rolled out across South Africa since November 2017. Since then it has evolved into a movement called Tangible Africa, as a collaboration between Computing Sciences and the NPO Leva Foundation. Three coding apps (TANKS, RANGERS and BOATS) are supplemented by coding kits, lesson plans and teacher training material [13]. Through various collaborations with NGO's, schools, teacher unions, libraries and other entities, the project has reached over 100 000 learners and trained over 20 000 teachers.

Since the planned Coding and Robotics curriculum has not yet been finalised, schools find it challenging to incorporate any coding activities within an already busy daily schedule. The following alternative ways have been identified by Tangible Africa [28]:

- Incorporating coding into other subjects – this is specifically successful in Mathematics, where various teachers have provided anecdotal proof that learners perform better in Mathematics once introduced to coding;
- Coding clubs – many schools have introduced voluntary coding clubs, where learners are introduced to various coding activities in the afternoon. This has proven to be very popular;
- Coding Ambassadors – since teachers are often overburdened by day-to-day curriculum work, they often do not have time to do additional activities. Consequently, Tangible Africa has rolled out 170 government sponsored interns into various communities across the country; and
- Coding tournaments – a great catalyst for learners and teachers to be involved in the coding project, is the presentation of coding tournaments. In Gqeberha, 5 schools have formed a coding league and once a term they compete for a floating trophy. The biggest annual event is the Mandela Day Tournament. In 2022, 6000 learners participated at 50 sites across all nine provinces.

Regarding the context of this paper, it is important to focus on the role Tangible Africa is playing. Within the Fourth Industrial Revolution, there is a growing shortage of software developer skills in South Africa as well as most countries in the world. As a result, the South African government is working towards the roll out of a Coding and Robotics curriculum in all primary schools. Major challenges that face this roll out include the following generic challenges across Africa [29, 30]:

- Lack of computers in schools;
- Electricity;
- Internet;
- Security;
- Teacher qualifications;
- Teacher intimidated by computers; and
- High cost of computers.

In South Africa, data regarding the availability of computer laboratories is not readily available. The latest data is from 2018 when it was reported that 16000 out of 25000 schools do not have computer labs [31]. At that stage it was estimated that it would cost government on average R1million per school to provide a connected computer lab. It is common knowledge that electricity supply and security is a growing challenge across all sectors of the country.

As a result of millions of learners consequently not being exposed to computers, many are unaware of the careers related to computing such as software development. They are thus effectively not given the opportunity to make career choices in this lucrative field. Two other realities in the South African schooling system makes this situation worse; the lack of career counselling at schools and the decreasing number of learners who are passing well in Mathematics.

Sefotho [32] found that the lack of career advice provided in South Africa is putting a great limitation on the youth. Especially learners in less-resourced communities who receive very little career counselling. News24 [22] quotes subject choice and career guidance counsellor, Shirley Brooks, as follows: "Sometimes, young people embark on further studies to fulfil their parents' dreams or because they are familiar with the particular direction and do not know what other options are available.

Cosser [33] describes the 2021 matric results in Mathematics as a national disaster, with less than 25% of Grade 12 scholars getting a mark of more than 50%. Keeping in mind what applicants generally need to be accept into a BSc Computer Science, it is even worse when noted that 13.1% achieved a mark of over 60%.

Facing these realities, Tangible Africa has set up the following Road Map to direct its impact on learners:

- Let's play – learners are introduced to the different coding apps, making them not only aware of coding, but also to careers in software development;
- Train the Teachers – the training material is designed for "low entry barriers" which implies that the vast majority of the 20 000 teachers experience no intimidation during training, and often leave feeling encouraged to start something at their schools. These training sessions also include some awareness about careers in computing;
- Tangible Academies – by rolling out the programme in schools, teachers identify learners with computational skills. Through the implementation of workshops on Saturdays, these learners are empowered mainly to achieve in Mathematics, as it is seen as critical for future computing careers. Where resources allow, learners are also provided with scholarships to attend good schools, which preferably teach IT as a subject; and
- Careers – where possible, Tangible Africa guides and supports matriculants to register for relevant computing degrees. At this early stage of the project, one student that has gone through the whole road map and is now employed as a software developer.

The Tangible Africa project thus addresses most of the challenges identified as resulting in learners not choosing computing careers.

6 Conclusions, Limitations and Future Research

Vernon and Drane [6] found that research informs influencers as to the importance of timely career pathway discussions. Recent research indicates that exposure to new technologies and being taught programming concepts at school level using mobile technologies can influence a scholar's IT career choice [19, 35, 36]. Social media was shown as an important influencer for first year students speaking an African language at home. The sources of information that first-year CS/IS/IT students identified in this study as useful were the university website, social media and respondents having an African home language being more influenced by IT role models.

Students from low socio-economic status back-grounds reported low expectations of attending university [6]. Research has indicated that parents, teachers and friends are key influencers for scholars' decisions in choosing an IT career [3, 23, 34, 35]. The findings of this study contradict general research findings that indicate the importance of parents, family and teachers as influencers of IT career choice in a developing country.

The research study has given insight into first year students' career influencers. The study has that research on career choice influencers, such as parents and teachers do not apply in the South African HEI environment. Thirty-five percent of the respondents indicated that their father was unknown and 19% do not know their mother. Finally, the study proposes that exposure to programming concepts, such as Tanks has a higher impact on influencing a child's IT career choice than parents.

The study recommends that academic departments understand the context of the potential student body and that assumptions are not made. Influencing future student's needs to be innovative with an understanding of their environments are indicated in the Social Learning Theory of Career Decision Making, Social Influence Theory and Latane's Social Impact Theory. The limitations of this study were the small sample size of CS/IS/IT first-year students and that the study was completed at one comprehensive university. Future research will investigate the impact of the Tanks project on first-year student's career choices and include an action plan on how to incorporate innovative initiatives like Tangible Africa in creating awareness of IT career options.

Acknowledgement. The paper is based on post-graduate research conducted by Twani [37] in the Department of Computing Sciences at the Nelson Mandela University.

References

1. Akosah-Twumasi, P., Emeto, T.I., Lindsay, D., Tsey, K., Malau-Aduli, B.S.: A systematic review of factors that influence youths career choices—the role of culture. Front. Educ. **3**, 58 (2018). https://doi.org/10.3389/feduc.2018.00058
2. Siddiky, R., Akter, S.: The students' career choice and job preparedness strategies: a social environmental perspective. Int. J. Eval. Res. Educ. **10**(2), 421–431 (2022). https://doi.org/10.11591/ijere.v10i2.21086
3. Abe, E.N., Chikoko, V.: Exploring the factors that influence the career decision of STEM students at a university in South Africa. Int. J. STEM Educ. **7**(1), 1–14 (2020). https://doi.org/10.1186/s40594-020-00256-x
4. Braza, M.R.S., Guillo, R.M.: Social-Demographic characteristics and career choices of private secondary school students. Asia Pac. J. Multidiscip. Res. **3**(4), 78–84 (2015)
5. Mtemeri, J.: Factors Influencing the Choice of Career Pathways among High School Students in Midlands Province. University of South Africa, Zimbabwe. PhD (2017)
6. Vernon, L., Drane, C.: Influencers: the importance of discussions with parents, teachers and friends to support vocational and university pathways. Int. J. Training Res. **18**(2), 155–173 (2021). https://doi.org/10.1080/14480220.2020.1864442
7. Kelman, H.C.: Compliance, identification, and internalization: three processes of attitude change. J. Conflict Resolut. **2**(March), 51–60 (1958)

8. Latané, B.: The psychology of social impact. Am. Psychol. **36**(4), 343–356 (1981). https://doi.org/10.1037/0003-066X.36.4.343

9. Krumboltz, J.D.: Improving career development theory from a social learning perspective. In: Savickas, M.L., Lent, R.W. (eds.) Convergence in Career Development Theories, pp. 9–32. Consulting Psychologists Press, Palo Alto, CA (1994)

10. Ireh, M.: Career development theories and their implications for high school career guidance and counselling. The High School J. **83**(2), 24–40 (2000)

11. Lee, P.C., Lee, J.M., Dopson, L.R.: Who influences college students' career choices? an empirical study of hospitality management students who influences college students' career choices? an empirical study of hospitality management students. J. Hosp. Tourism Educ. **31**(2), 74–86 (2019). https://doi.org/10.1080/10963758.2018.1485497

12. Kapoor, A., McCune-Gardner, C.: Understanding CS undergraduate students' professional identity through the lens of their professional development. In: Proceedings of 24th Annual ACM Conference on Innovation and Technology in Computer Education, pp. 9–15 (2019)

13. Greyling, J.: Coding unplugged—a guide to introducing coding and robotics to South African schools. In: Halberstadt, J., de Bronstein, A.A., Greyling, J., Bissett, S. (eds.) Transforming Entrepreneurship Education: Interdisciplinary Insights on Innovative Methods and Formats, pp. 155–174. Springer International Publishing, Cham (2023). https://doi.org/10.1007/978-3-031-11578-3_9

14. Calitz, A.P, Greyling, J.H., Cullen, M.D.M.: ICT career track awareness amongst ICT graduates. In: Proceedings of the South African Institute of Computer Scientists and Information Technologists Conference on Knowledge, Innovation and Leadership in a Diverse, Multidisciplinary Environment, pp. 59–68 (2011), https://doi.org/10.1145/2072221.2072229

15. Ward-Smith, P.: Issues: the fine print of literature reviews. Urol. Nurs. **36**(5), 253 (2016). https://doi.org/10.7257/1053-816X.2016.36.5.253

16. Kitchenham, B., et al.: Guidelines for performing Systematic Literature Review of Software Engineering – Technical Report (2007)

17. Matthew, G., Owusu, Y., Bekoe, R.A.: What influences the course major decision of accounting and non-accounting students ? J. Interact. Online Learn. **12**(1), 26–42 (2018). https://doi.org/10.1108/JIEB-02-2018-0004

18. Bock, S.J., Taylor, L.J., Phillips, Z.E., Sun, W.: Women and minorities in computer science majors: results on barriers from interviews and a survey. Issues Info. Sys. **14**(1), 143–152 (2013). https://washburn.edu/academics/college-schools/arts-sciences/departments/computer-information-Sciences/files/BockEtal2013.pdf

19. Aivaloglou, E., Hermans, F.: Early Programming Education and Career Orientation: The Effects of Gender, Self-Efficacy, Motivation and Stereotypes. In: Proceedings of the 50th ACM Technical Symposium on Computer Science Education (SIGCSE '19), 27 Feb–2 March, Minneapolis, MN, USA, pp. 679–685 (2019). https://doi.org/10.1145/3287324.3287358

20. Alvarado, S.E., An, B.P.: Race, friends, and college readiness: evidence from the high school longitudinal study. Race Soc. Probl. **7**(2), 150–167 (2015). https://doi.org/10.1007/s12552-015-9146-5

21. Mein, E., Esquinca, A., Monarrez, A., Saldaña, C.: Building a pathway to engineering: the influence of family and teachers among mexican-origin undergraduate engineering students. Hispanic Educ. (2018). https://doi.org/10.1177/1538192718772082

22. Hako, N.: 'Young and uninformed': Why South Africa's youth desperately needs career guidance. News24 (2021). https://www.news24.com/life/archive/young-and-uninformed-why-south-africas-youth-desperately-needs-career-guidance-20210720

23. Säde, M., Suviste, R., Luik, P., Tõnisson, E., Lepp, M.: Factors that influence students' motivation and perceptions of studying computer science. In: Proceedings of the 50th ACM Technical Symposium on Computer Science Education (SIGCSE'19) (2019), https://doi.org/10.1145/3287324.3287395

24. Downes, T., Looker, D.: Factors that influence students' plans to take computing and information technology subjects in senior secondary school. Comp. Sc. Educ. **21**(2), 175–199 (2011). https://doi.org/10.1080/08993408.2011.579811

25. Govender, I., Khumalo, S.: Reasoned action analysis theory as a vehicle to explore female students' intention to major in information systems. J. Comm. **5**(1), 35–44 (2014)

26. Seymour, L.F., Serumola, T.: Events that lead university students to change their major to Information Systems: A retroductive South African case. SA Comp. J. **28**(1), 18–43 (2016). https://doi.org/10.18489/sacj.v28i1.367

27. Batteson, B.: Investigation and development of an inexpensive educational tool suite for an introduction to programming. Honours Treatise, Nelson Mandela University (2017)

28. Greyling, J.H.: Guidelines for Introducing Learners to Computer Programming in a Developing Country, CT Education: 51st Annual Conference of the Southern African Computer Lecturers' Association, SACLA 2022, Cape Town, South Africa, 21–22 July 2022. Revised Selected Papers (2022)

29. ICT Works: 12 Challenges facing ICT Education in Kenya, 12 Sep 2011. https://www.ict works.org/12-challenges-facing-computer-education-kenyan-schools/

30. The Conversation.: Coding in South African schools: what needs to happen to make it work (2019). https://theconversation.com/coding-in-south-african-schools-what-needs-to-happen-to-make-it-work-120861

31. BusinessTech.: Here's how many South African schools don't have the Internet or a computer lab—and what it will cost to fix the problem. 18 Jul 2018, https://businesstech.co.za/news/int ernet/259171/heres-how-many-south-african-schools-dont-have-the-internet-or-a-computer-lab-and-what-it-will-cost-to-fix-the-problem/

32. Sefotho, M.M.: Career guidance in South Africa as a social justice travesty. Orientación y Sociedad **17**, 153–163 (2017)

33. Cosser, M.: Daring solutions are needed to solve South Africa's maths teaching crisis. Daily Maverick (2023), https://www.dailymaverick.co.za/article/2023-01-22-daring-solutions-are-needed-to-solve-south-africas-maths-teaching-crisis/

34. Stone, J.A.: Student perceptions of computing and computing majors. J. Comput. Sci. Coll. **34**(3), 22–30 (2019)

35. Potter, L.E.C., von Hellens, L.A., Nielsen, S.H.: Childhood Interest in IT and the Choice of IT as a Career: The Experiences of a Group of IT Professionals. SIGMIS-CPR'09, 28–30 May, Limerick, Ireland (2009)

36. Mano, L.: TANKS: An app that teaches coding, without a computer. ITWeb TechForum (2023). https://www.itweb.co.za/content/xnklOvzbQnpv4Ymz

37. Twani, M.: Factors Influencing First-Year Students' Career Decisions to Pursue An IT Career. MSc Degree in Computer Science and Information Systems, Nelson Mandela University, South Africa (2021)

Author Index

H. E. Van Rensburg et al. (Eds.): SACLA 2023, CCIS 1862, p. 181, 2024.
https://doi.org/10.1007/978-3-031-48536-7